Successfully Installing TPM *in a Non-Japanese Plant*

Total Productive Maintenance

Successfully Installing

TPM

in a Non-Japanese Plant

Edward H. Hartmann, P.E.

TPM Press, Inc.
Pittsburgh, Pennsylvania   Charlotte, North Carolina

 Additional copies of this book are available from the publisher. Address all inquiries to:

TPM Press, Inc.
4018 Letort Lane
Allison Park, PA 15101, USA
Phone: (412) 486-6340
Fax: (412) 486-6375

Library of Congress Catalog Card Number: 92-39939
ISBN: 1-882258-00-2

Cover Design by Harold Behar
Printed and bound by Delmar, a division of
Continental Graphics Corporation
Printed in the United States of America

Library of Congress Cataloging-in-Publication Data

Hartmann, Edward.
Successfully installing TPM in a non-Japanese Plant: total productive maintenance / Edward H. Hartmann.
Includes index.
1. Total productive maintenance 2. Plant maintenance - Management
3. Industrial equipment - Maintenance and repair - Management
I. Title. II. Title: Total productive maintenance.
TS192.H367 1992 658.2'02--dc20 92-39939
ISBN 1-882258-00-2 CIP

10 9 8 7 6 5 4 3

Table of Contents

List of Figures ix

Preface xi

I. The World of TPM 1
The Japanese and TPM 2
The Will and the Way 3
Meeting the Challenge 5

II. Meeting the Manufacturing Challenge 7
Global Competition 7
The Quality Challenge 8
Just-In-Time (JIT) 9
Cycle Time Reduction 9
Set-Up Reduction 10
Cost Reduction 10
Capacity Expansion 11
Other Issues 12
The TPM Solution 13

III. Equipment, the Focus of TPM **15**
TPEM (Total Productive Equipment Management) 16
Asset Utilization 16
Equipment Management 17
Lowering the Life Cycle Cost 25
Getting the Best from Your Machines 29
TPM Goals 31
The Three Zeros 32
The Elements of TPEM 34
The Meaning of Total 35
The TPM Organization 37

IV. The Power of TPM **39**
TPM Impacts all of Manufacturing 40
Reducing Defects 41
The Passion of Productivity 42
Controlling Maintenance Costs 42
Improving Your Safety Record 45
The Bottom Line 45
Employee Participation 47
Putting the Power of TPM to Work 48

V. Measuring Your True Equipment Productivity **51**
Unlocking the Hidden Factory 51
Equipment Productivity 52
Equipment Losses 54
Calculating Equipment Effectiveness 59
Applying the Formulas 64
OEE Goals 66
Setting Your Priorities 68

VI.	**Customizing Your TPM Installation**	**69**
	TPM Strategy	69
	The Components of TPEM	70
	Autonomous Maintenance with a Difference	71
	An Ounce of Prevention is not Enough	74
	Improving Your Equipment	75
	TPM Installation Strategies	76
VII.	**How Much Autonomous Maintenance Do You Need?**	**79**
	Custom-Made Autonomous Maintenance	80
	Recognizing Limitations	81
	Training--The Key to TPM-AM	83
	On-the-Job Training	84
	Levels of Training	86
	The Cost of TPM-AM	86
	Certification	88
	"My Machine" Concept	90
VIII.	**How to Design and Install an Effective PM Program**	**93**
	Types of PM	95
	PM Strategy	98
	An Effective PM System	98
	The Secrets of Successful PM	111
	Operator-Based PM	112
	Computer Friendly	113
IX.	**Improving Equipment Through Problem-Solving Techniques**	**115**
	CATS	116
	Feed the Cats	117

CATS Meetings 127
Analyzing the Problems 128
Reaction of Operators 130

X. The Feasibility Study 133
Contents of a Feasibility Study 135
Organization of the Feasibility Study 149
Execution of the Feasibility Study 150
Feasibility Study Report and Presentation 153

XI. The TPM Installation 157
Phase I: Installation Planning and Preparation 157
Phase II: Pilot Installation 181
Form Teams 182
Installing TPM-EM 183
Installing TPM-PM 190
Installing TPM-AM 194
Phase III: Plant-Wide Installation 200
Training 201
Maintenance Management System 202
Reporting Progress 203
Recognition, Reinforcement and Celebration 204

XII. "Fast-Track" TPM Installation 207

About the Author 215

Index 217

List of Figures

Figure 1	Phase I Equipment Management	19
Figure 2	Phase II Equipment Management	24
Figure 3	Life Cycle Cost (LCC)	27
Figure 4	Phase III Equipment Management	30
Figure 5	Typical TPM Organization	37
Figure 6	Maintenance Department from "Low Tech" to "High Tech"	44
Figure 7	The Three Major TPM Formulas	53
Figure 8	Equipment Losses	55
Figure 9	Definitions for Equipment Loss Calculations	60
Figure 10	Calculation Example	61
Figure 11	Equipment Productivity and Performance Calculation	65
Figure 12	The Cumulative (Cubed) Effect of TPM on Equipment Effectiveness	67
Figure 13	Components of TPEM	70
Figure 14	Operator Training under TPM-AM	72
Figure 15	Chart of Skill Levels-Operators	87
Figure 16	Average Operator Skill Level	89

Figure 17	TPM Equipment History	104
Figure 18	TPM Small Group Activities	117
Figure 19	Pareto Analysis	119
Figure 20	TPM Equipment Condition Analysis	121
Figure 21	Condition-Action	123
Figure 22	Failure Information Sheet (FISH)	125
Figure 23	Cause and Effect Diagram (Fishbone)	129
Figure 24	OEE Observation and Calculation Form	137
Figure 25	Skills Required/Available Analysis	140
Figure 26	Current Maintenance Assessment	144
Figure 27	Feasibility Study Schedule	152
Figure 28	TPM Development, Flow & Organizations	161
Figure 29	TPM "Job Descriptions"	163
Figure 30	Policy Letter No.18 (Ford)	166
Figure 31	TEC Newsletter Article	171
Figure 32	Ford Enfield/Treforest Newsletter	174
Figure 33	TPM Master Plan	176
Figure 34	Pilot Installation Plan	178
Figure 35	Eight-Step Method to Establish Priorities	185
Figure 36	OEE Improvement Guide	189
Figure 37	PM Task Transfer Analysis	191
Figure 38	Initial Cleaning Schedule	195
Figure 39	Action Plan Based on Initial Cleaning	197
Figure 40	Equipment Checklist	199
Figure 41	Recognizing Equipment Performance	205
Figure 42	A-B-C Analysis of TPM Activities	209
Figure 43	"Fast-track" Installation Schedule	211

Preface

This book was written for the plant manager, the production manager, the maintenance manager, the TPM manager and all other managers, engineers and supervisors who are faced with implementing TPM. Before becoming interested in TPM, I was a consultant in the areas of maintenance management and maintenance productivity, which focuses on productivity improvement of the maintenance department; i.e. people. TPM focuses on the improvement and performance of *equipment*.

In 1986, I went to Tokyo to visit with Seiichi Nakajima, the acknowledged "Father of TPM," and with his kind help, was able to visit and study numerous small and large, but all highly successful TPM installations in Japan. It became clear to me that TPM is a sorely needed technology that enables any manufacturing plant to dramatically improve the productivity of its equipment and its state of maintenance. In 1987, Mr. Nakajima and I introduced TPM to American industry during a press conference and executive meeting especially called for that purpose. Since that time, I have been involved with TPM training and consulting in many companies around the world.

Unfortunately, part of that activity included working with plants that had started a TPM implementation that did not go well. In all instances, no feasibility study had been carried out and little thought was given of the need to adapt the process to a different culture and local circumstances. It became necessary to develop a different approach that enables a non-Japanese plant to successfully install TPM within their own particular environment and suited to their personnel, without giving up any of the benefits that TPM offers.

This book discusses an approach and process that has been used successfully in plants in Asia, Europe, Latin America and North America. I sincerely hope it will help you to develop and install a TPM program in your plant that achieves world-class manufacturing excellence in a proud, productive and pleasant environment.

Edward H. Hartmann, P.E.
President
International TPM Institute, Inc.

Successfully Installing TPM *in a Non-Japanese Plant*

CHAPTER I

The World of TPM

Total Productive Maintenance--TPM--anywhere you go in the world today, maintenance, production and plant managers are talking about it, trying to find out more about it, or learning how to implement it. From electronics factories in Malaysia to automotive companies in the United States and Europe to aluminum process industries in Canada and paper plants in South America, *everyone* is turning to TPM as the latest and best in the series of modern manufacturing techniques. Companies like Ford Motor, Motorola, Eastman Kodak, DuPont, Texas Instruments, Procter & Gamble, IBM, AT&T, and many others have begun to install TPM programs in their plants or already have successful installations. It truly seems to be one of the important waves of the future in manufacturing technology.

But is TPM really the solution to your problems? Can it help you improve your plant operations, increase your productivity, and reduce costs? The answer is a resounding yes, but it must be done the *right* way that produces results in *your* plant. TPM has produced tremendous benefits in the U.S. and other countries but may not work for you if you simply try to copy the Japanese system. Some companies outside Japan have not accomplished good results and experienced a great deal of frustration and delays by simply following the Japanese model. Total Productive Maintenance will be most effective if you modify it to work for you, in your environment, with your people, addressing your equipment problems and opportunities.

This book will take you step by step through a TPM process that is designed to produce results in *your* plant. It points out problems others have had during their TPM development and should help you implement TPM successfully in your plant.

The Japanese and TPM

The planned approach to *Preventive Maintenance* (PM) was introduced into Japan not long after its invention by General Electric Corporation in the 1950s. As has happened with so many other productivity techniques, such as Quality Management and Zero Defects, the Japanese took the concept and developed it into an improved program that helped them become more efficient producers. Seiichi Nakajima, Vice Chairman of the Japan Institute of Plant Maintenance, promoted TPM throughout Japan and became known as the "Father of TPM." He wrote a book on the subject, which has become the Bible for maintenance and production managers in Japan and many other countries.

Nakajima's approach, first practiced by Nippondenso, to include the operators in routine maintenance activities is sound, and forms the basis for good TPM. But Nakajima's text was written with Japanese companies in mind. TPM has been recognized and reported in many management journals to be one of the strongest tools the Japanese have been using over the last 20 years to gain a manufacturing advantage over the rest of the world.

There is no doubt that the Japanese have a head start in this area. They also have other advantages. There is total corporate management commitment to TPM in Japan. The president or chairman of the board of a large Japanese corporation can tell one of his supplier companies that if they want to continue selling to his firm, they should install TPM. The president of that supplier company goes back to his plant and informs his employees that they will start a TPM program. His subordinates don't argue about it, no one has second thoughts about it; they just pursue this goal with all the resources at their disposal.

The Will and the Way

Speaking of resources, when the Japanese decide to pursue a certain goal, they make sure it is adequately funded. There is no holding back on the yen, and there is total management commitment. You will not always find this attitude in the U.S. or other Western countries. Often, it is a middle manager who first discovers the benefits of TPM. He or she must then go back and sell top management, and, in addition, justify the cost of the program.

The Japanese do very long range planning. They routinely plan ten and twenty years ahead. If a TPM program is going to take three or more years to show results, they have the patience and the commitment to make it happen. In the U.S. and other countries, the objectives are usually short range. If you can't show a profit in six months to a year, your program is in jeopardy of being canceled.

Many Japanese, especially middle managers, work a ten-hour day or more. In addition, many employees have a two hour or even longer commute each day. And on top of that, they volunteer additional hours to learn and practice TPM. If you tried to add TPM training on a voluntary, non-paid overtime basis in the U.S. or other Western countries, the employees would resist.

Another aspect of the Japanese culture is their natural affinity for groups. They are trained that way from childhood. In the schools, the children form teams to serve lunch. Groups of children clean the floors and the schoolyard. The Japanese children must participate in team activities after school, whether it is playing an instrument, participating in a sport, or acting in theater. These children grow up in groups, which makes it easy, almost natural for them to join the small groups that form the heart of TPM. Americans and other nationalities tend to be more independent, and act in their own interests, rather than those of a group.

For you to succeed with TPM, you must first be aware of these differences in work ethic, management style and cultural background. You cannot copy the Japanese, because many who have tried have not been very successful. You must be pragmatic and develop a program that works for *you*, in *your* environment, with *your* people. You *can* develop a customized TPM process that will produce the same excellent results in your plant as the Japanese have accomplished in theirs.

Meeting the Challenge

It's a formidable challenge, but one well worth the time, effort and money. You have to get top management support, which will come 90% of the time, once they clearly understand what TPM is and what it can do for their operations. If you have a union, make sure that they are involved from the very beginning. You don't need their enthusiastic support, but you must at least have their participation and acceptance.

Before you embark on a TPM program, you must be aware that it isn't going to be free. In fact, it will take a sizeable investment of time, money, and training effort before you begin to see substantial results. You must also realize that you can't rush into TPM unprepared. It's a drastic step that will change your corporate culture and will take a lot of preparation and strong commitment.

With these cautions, you are ready to develop a workable plan to improve your plant by installing a proven system to run your equipment with a minimum of downtime and greatly increased productivity.

CHAPTER II

Meeting the Manufacturing Challenges

Every company faces a continuous challenge to improve their operations and their way of doing business. Standing still means falling behind. What was acceptable five years ago will put you at a disadvantage today. The best companies, those who will be around in the future, are continuously improving and renewing themselves.

TPM will address many manufacturing challenges that may be confronting you. A short review of the major issues will be useful in understanding how TPM can help to overcome these challenges.

Global Competition

The competition your company faces is fierce and typically global in scope. You're competing in international markets even if you don't export, because chances are someone is importing a similar product to compete with you in your own backyard. Even different plants within the same corporation must compete for new products. As a result, you must promise your customer total satisfaction. And you must deliver on that promise!

The Quality Challenge

Motorola and others have announced that their quality goal is Six Sigma. That means 99.9996% good parts delivered to their customer. Put in production terms, that's a rejection rate of 3.4 parts per million (ppm). In other words, you must make 300,000 good parts before you ship a bad one! If your plant makes a concentrated effort to attain this level of quality, you should know in advance that at a certain point, like the marathon runner, you will hit a wall, a barrier that will seem to frustrate your efforts.

What is this obstacle? It's your machines. You must have a perfect machine to produce a perfect product. Even the best machines can go downhill quickly if they are not properly maintained. A paper company recently installed a brand new papermaking machine which began having many breakdowns after only nine months. On inspection, it was found that the machine was already rusting badly, was barely cleaned and received practically no maintenance. For perfect quality, you need the perfect machine, something that many people don't believe can exist. An absolutely clean, perfectly maintained, totally-in-adjustment machine with no worn parts? Impossible!

But Japanese TPM companies are doing it with their

machines. Their factories are sparkling clean and they have a mania for cleanliness and continuous maintenance. Can it be done in the West? The critics may scoff, but all it takes is continuous nonstop attention to your machines. You can get this commitment through Total Productive Maintenance.

If you already have a successful quality program, it will make your TPM installation much easier. In fact, quality and TPM fit well together, and you won't have to create a totally new program. TPM programs have been very successfully installed as part of an overall quality program -- quality of equipment. If you don't currently have a quality program in your plant, you can still have a good TPM installation, which will result in improved quality of your product.

Just-In Time (JIT)

Another modern production technique is just-in-time. It is a highly efficient method of operation that reduces your inventory levels considerably, both in-process and finished goods. But JIT depends on reliable equipment. If you have an equipment breakdown in the middle of a JIT run, you immediately wipe out all the gains you have made.

Cycle Time Reduction

State-of-the-industry factories present you with another challenge. More and more, you must be able to make faster production runs to reduce cycle time. Satisfying the customer means your production runs become shorter and shorter to produce customer orders with less lead time. Equipment breakdowns, idling and minor stoppages will make it very difficult to reduce cycle times if they are not systematically addressed by your TPM program.

Set-Up Reduction

JIT and cycle time reduction result in typically shorter and more frequent production runs. Now, suddenly, your set-ups and changeovers become crucial. Because during a set-up or a changeover, your machine is *down*. It is not *broken down*, but nevertheless, it is *down* as far as running any production is concerned. Past Overall Equipment Effectiveness (OEE) studies have shown that set-ups and adjustments can consume up to 50% of total production time. Set-up and adjustments is one of the recognized major equipment losses under TPM.

Single Minute Exchange of Die (SMED) is a very successful approach to reduce set-up time to an absolute minimum. There are many examples in industry, where a set-up that took one and a half hours has been reduced to 45 minutes and then to ten minutes. The goal of SMED is a *single digit* number of minutes, meaning less than ten for *any* set-up! Under TPM, the involvement of the operators in accomplishing reduced set-up times becomes very important and has shown dramatic results.

Cost Reduction

The attention of past cost-reduction efforts has mostly been on manufacturing costs. However, maintenance costs typically make up between 5-15% of total production costs. The actual number depends on the type of company measured. Heavy process industries will fall toward the high end of this scale, while industries with a large amount of manual labor and few machines will be at the low end. Highly automated plants experience a much higher percentage than cited above.

The critical indicator is not only the actual cost, but the trend. Production costs per unit have been coming down over time due to automation, faster equipment, robotics, cost reduction studies and more. Maintenance costs, on the other hand, have been escalating. As equipment becomes more complicated and sophisticated, the cost of maintenance goes up. The focus directed only on production costs is now changing. Many companies are actively searching for ways to reduce maintenance costs. But if you perform primarily breakdown maintenance, which you cannot control or predict, how can you reduce maintenance costs? A good TPM installation will reverse your spiraling maintenance costs and greatly improve your equipment's performance at the same time.

Capacity Expansion

Manufacturing produces a product. Maintenance creates the capacity for production. Do you know the real *utilization* of your equipment? Do you know your real equipment *availability*, when the equipment is being utilized? Do you know your real equipment *performance* when it is utilized and available?

Careful equipment studies have shown some incredibly low Total Effective Equipment Productivity (TEEP) numbers, much to the surprise and consternation of management. Not only on old and worn out equipment, but sometimes on rather new and modern machines! Not only on unimportant, redundant equipment, but sometimes on production-limiting constraint equipment, reducing total throughput!

Sometimes, there is so much available capacity hidden in your existing equipment, that you could postpone planned equipment purchases or even a plant expansion for years by simply learning how to tap that unused capacity. Some companies, such as Tennessee Eastman, have clearly demonstrated that TPM can, sometimes dramatically, increase production capacity without capital investment.

Other Issues

Environmental issues are becoming a major factor in many countries. The regulations are getting tougher and tougher every month. Your machines must not pollute the air, ground, or water, yet must run faster and produce more. Only well maintained and properly adjusted and inspected equipment can cope with these issues.

The other side of the environmental coin is energy conservation. Electrical motors are the highest energy consumers in most manufacturing plants, yet many electrical motors run at a low efficiency, due to partially burnt windings, bad insulation, dirt accumulation or imbalance. The challenge is to reduce energy consumption. How can you handle this demand and still get more capacity from your equipment?

The TPM Solution

These are some of the problems and challenges your company has to face to remain competitive. Total Productive Maintenance, properly installed, has a positive and often dramatic effect on many of these issues without you paying an exorbitant price for your quality and productivity gains. Quite the opposite, the return on investment (ROI) of your successful TPM installation is likely to be higher than any of your previous productivity improvement programs.

CHAPTER III

Equipment, the Focus of TPM

What exactly is TPM? In Japan, it is often defined as "productive maintenance involving total participation." Part of the complete definition includes, in addition to maximizing equipment effectiveness and establishing a thorough system of PM, a statement that "TPM involves every single employee."

This definition is accurate, of course, but it is the Japanese approach. It focuses on "maintenance" and on "every single employee", a notion that has caused problems in many non-Japanese companies. A more appropriate and acceptable Western approach concentrates on the machine. Hartmann's definition of TPM as practiced by Western companies is:

"Total Productive Maintenance permanently improves the overall effectiveness of *equipment*, with the active involvement of operators."

This definition puts the emphasis on "overall effectiveness of equipment" and not on maintenance and on "active involvement of operators" instead of "every single employee." While TPM involves more than maintenance personnel and operators, such as engineering, purchasing, supervision and others, the gains in overall equipment effectiveness are clearly accomplished by good teamwork between operators and maintenance crafts.

TPEM (Total Productive Equipment Management)

The new *process*, developed by International TPM Institute, Inc., that makes it easier to accomplish a successful, tailor-made TPM installation in a non-Japanese plant, is called Total Productive Equipment Management (TPEM).

Unlike the rather formalized, rigid Japanese TPM program, TPEM allows you to develop a highly flexible installation. It takes into consideration your actual equipment needs and priorities and in particular your specific corporate and plant culture (especially if you are unionized). This is the *pragmatic* versus what could be called a *dogmatic* approach.

Asset Utilization

Next to land and buildings, production equipment is normally a manufacturing company's largest asset. Return on assets (ROA) is a widely used measure of financial performance. Asset *utilization* is the single most important factor that influences the Return on Assets results.

However, equipment utilization, equipment availability and equipment performance are often found to be exceedingly low, resulting in *very low* asset utilization.

Therefore, one of the most important considerations when developing your TPM installation is improved equipment management, to improve asset utilization.

Equipment Management

TPM, through the TPEM (Total Productive Equipment Management) process, will re-focus and restructure your company's approach to equipment management. In addition to *equipment utilization* (running it a high percentage of a 24 hour day), *equipment performance* and *equipment availability* are the *key ingredients* to sound equipment management and high asset utilization.

The three phases of improving equipment management for most companies are:

I Improvement of existing equipment

II Maintaining improved (or new) equipment at a higher level of performance and availability

III Procurement of new equipment with high level of performance and availability

In a new plant, just receiving equipment, Phase I is replaced with "Buy-off of equipment only when a specified level of performance has been accomplished" and/or "Debugging of equipment to accomplish a specified level of performance."

Each phase of Equipment Management contains several steps, which must be carefully considered when designing the TPM installation for your plant.

The first phase of Total Productive Equipment Management is to improve your equipment to its highest required level of performance and availability (see Figure 1). This is a major and most important phase of TPM. Depending on the current condition and performance of your equipment, this phase may take a long time and a fair amount of money and effort. However, major gains in productivity, quality improvement and cost reductions are accomplished here.

It is very important to approach this phase with sufficient valid data and careful planning. You should establish a list of priorities in order to first accomplish improvements on your constraint equipment for quick improvement of your throughput. This approach will also accomplish an early break-even point, when benefits exceed costs.

The first three steps develop the data needed for decision making and setting of priorities. Carry out these steps as part of your feasibility study (Chapter X). Frequently, the final management decision to proceed with TPM is made after the feasibility study. Using the input from this study and other data (such as existing breakdown records, failure information

EQUIPMENT MANAGEMENT

PHASE I

IMPROVE EQUIPMENT TO ITS
HIGHEST REQUIRED LEVEL OF PERFORMANCE
AND AVAILABILITY

Step 1. Determine existing equipment performance and availability (current OEE)

Step 2. Determine equipment condition

Step 3. Determine current maintenance (especially PM) performed on equipment

Step 4. Analyze equipment losses

Step 5. Develop (and rank) equipment improvement needs and opportunities

Step 6. Develop set-up or change-over improvement needs and opportunities

Step 7. Execute improvements as planned & scheduled

Step 8. Check results and continue as required

Figure 1

sheets, equipment histories, repair costs and mean time between failures data) the TPM teams analyze equipment losses (step 4) and develop equipment improvement needs and opportunities (step 5). Cost-benefit analysis, throughput requirements, quality improvement needs, time availability and other considerations will determine the ranking of improvement projects.

Step 6 of Phase I addresses set-up or changeover improvement needs and opportunities. The same TPM small groups (maybe with specialized engineering support) analyze the set-up losses, develop improvement needs and design improvement projects. The next activity (step 7) is the actual execution of improvement projects according to schedule. Depending on the condition of your equipment and the established needs and opportunities, this step can take quite a long time (six to 18 months). It actually never really stops, as equipment needs continuous improvement. However, this step normally produces the quickest and most significant TPM results. It is one of the most exciting and profitable TPM activities, with significant impact on equipment performance, asset utilization, quality of product, throughput and cost. The major portion of "return" on investment is produced here, by the small groups, maintenance and engineering working in close cooperation.

The last step of equipment improvement (step 8) is to measure and publish the results as compared to baseline and to continue with equipment improvement activities as required.

Phase II of Equipment Management is to maintain equipment at its highest level of performance and availability. Here, you insure that the improvements made in Phase I don't evaporate. Or, if you have new equipment, you must make sure that a high level of performance is maintained throughout its entire useful life. The key point to remember here is that nothing can substitute for good preventive maintenance to accomplish this goal. Part of a good PM system is predictive maintenance (PDM), using state-of-the-art diagnostic equipment to forecast potential equipment failures and to fix these problems before they cause equipment failure.

Complicated and expensive diagnostic equipment is not always necessary to keep your machines in peak operating condition. Often all you need is careful inspection to expose hidden defects or to prevent potential problems.

Cleaning is another tool that helps to keep your machines running at peak efficiency and to improve product quality. It may not seem that important with all the other activities going on, but its effect on overall productivity can be dramatic.

A North Carolina cigarette plant offers an excellent example. In this plant, there are a large number of high-speed machines that make an average of 7,000 cigarettes a minute. In this highly automated mass production process, a continuous strip of paper is fed into the machine from a large roll and formed into a U-channel, where the tobacco is added. Further down, the paper is formed into an open O, where glue is applied to one side of the paper and the "rod" (an infinite length of cigarette) is formed. If the glue supply from a small tube is interrupted, the tobacco falls out and spills into the machine. The equipment must be shut down and cleaned, and the tip of the glue supply tube must be inspected and cleaned. In the next operation, the "rod" is cut into a double length cigarette and, after adding the filter, cut again into individual cigarettes. Each cutting process produces "shorts" (small pieces of tobacco), that can fall into the machine and also accumulate in other places, such as motor housings, electrical panels, on top of the equipment, in the cavities of rotating drums, etc. It does not take a lot of imagination to figure out how those particles, along with the dust of the charcoal filter inserts, can cause jams and other equipment stoppages.

Typically in such situations, line maintenance personnel (in this plant aptly called "stand-around" maintenance) are called to fix the problem and get the machines going again. However, due to the random nature of these breakdowns, maintenance is often busy at another machine, causing delays and excessive downtime. One group of operators in a somewhat remote area got tired of this situation and came up with their own solution. They knew the critical points in the machine where jamming and other problems (such as with the glue) often occurred. They began cleaning their machines on a regular basis and started to fix minor problems themselves.

As a result of their initiative, their machines needed less attention from the maintenance personnel, but produced up to 20% more output!

As this story illustrates, cleaning and inspection of equipment by the operators is one of the most powerful (and under-utilized) tools to keep machines running and to improve productivity and quality. Therefore, during Phase II of Equipment Management (Maintain Equipment at its highest Level of Performance), cleaning, lubrication and inspection activities by operators play a dominant role.

Step one of this phase (see Figure 2) is to develop the PM requirements for each machine. This is done by a team from engineering, maintenance and operators, based on experience and manufacturers recommendations. It includes PM activities that could be done by operators now or after training (Type I), and PM activities that will be done only by maintenance (Type II).

The same process is used to develop lubrication requirements for each machine (Step 2), as well as cleaning requirements (Step 3). The next step is to develop PM, lubrication and cleaning procedures, which will serve as a basis for training and for PM checklists, work orders and schedules. Step 5 is the development of inspection procedures for each machine. Usually, inspection activities are part of PM, but sometimes inspection is a separate activity to determine component wear and to find other potential problems. As with PM, cleaning and lubrication, there are Type I and Type II inspection activities (done by operators or by maintenance).

Step 6 is the development of the forms required to plan, execute and control all PM, lubrication, cleaning and inspection activities. Forms include checklists, work orders, schedules, inspection sheets, reports, etc. The development of the system takes place.

The next step is the development of a PM manual. It should include the PM philosophy within TPM, the plant-

EQUIPMENT MANAGEMENT

PHASE II

MAINTAIN EQUIPMENT AT ITS
HIGHEST REQUIRED LEVEL OF PERFORMANCE
AND AVAILABILITY

Step 1. Develop PM requirements for each machine

Step 2. Develop lubrication requirements for each machine

Step 3. Develop cleaning requirements for each machine

Step 4. Develop PM, lubrication and cleaning procedures

Step 5. Develop inspection procedure for each machine

Step 6. Develop the PM, lubrication, cleaning and inspection system, including all forms and controls

Step 7. Develop the PM manual

Step 8. Execute PM, cleaning and lubrication as planned and scheduled

Step 9. Check results and correct as required

Figure 2

wide PM policy, all the procedures for PM, lubrication and inspection activities, and the PM organization. Guidelines for development and use of the PM checklists, work orders, schedules and controls (including MTBF, costs, trends, etc.) are also part of a PM manual.

After these preparatory steps, the execution of the PM, cleaning, lubrication and inspection activities can begin. At this point, the operators start participating, depending on their level of skill and motivation. As time progresses, operators will and also want to assume more and more Type I activities within the TPM philosophy and policy of your plant. An organized process of skill transfer and training will take place.

Under step 9, the results (impact of improved PM, lubrication, cleaning, inspection) will be measured and the tasks and frequencies can be corrected as required. The most successful PM system is a dynamic approach that responds to the actual and sometimes changing condition of the equipment. It may even result in the reduction of activities or increase of time intervals between actions, if the condition of the equipment or the components is perfect at the time of PM. Active feedback from maintenance and operators is required to accomplish that.

Lowering the Life Cycle Cost

Phase III of TPEM is to procure (buy or build internally) new equipment with a defined level of high performance and low Life Cycle Cost (LCC). Simply defined, LCC contains all the costs incurred during the equipment's lifetime. There are five major stages which every piece of equipment goes through (Figure 3).

Equipment design comes first, and up to 80% of a machine's LCC is determined here (whether a machine is fully automatic or requires operators, how maintenance and repair-intensive a machine will be, etc.). Equipment construction comes next, after which, the machine is delivered. Third is equipment installation and debugging, which sometimes accounts for a high percentage of purchase cost, whether intended or not.

Fourth is the cost of operation (annual wages of operators and supervisors x number of years equipment is used, plus other operational costs). This is usually the largest LCC element for a machine that is run by operators over a number of years (the typical case). The final item is the cost for maintenance, repairs, rebuilds or overhauls and equipment improvements. Normally, the last two cost items far exceed the purchase price of the equipment. All five cost items are variable, and it is found that because of unwise equipment purchasing decisions, savings in acquisition costs (items 1-3) often result in large penalties when cost items 4 and 5 (operation and maintenance) accrue over the life of a machine.

Therefore, a new approach is needed that takes the *total* LCC picture into consideration. While up to 80% of LCC costs are determined at the design (or specification) stage, most of the LCC costs are *spent* on operation and maintenance of the equipment. How can you improve that situation and lower the LCC of your new equipment?

LCC

(Life Cycle Cost)

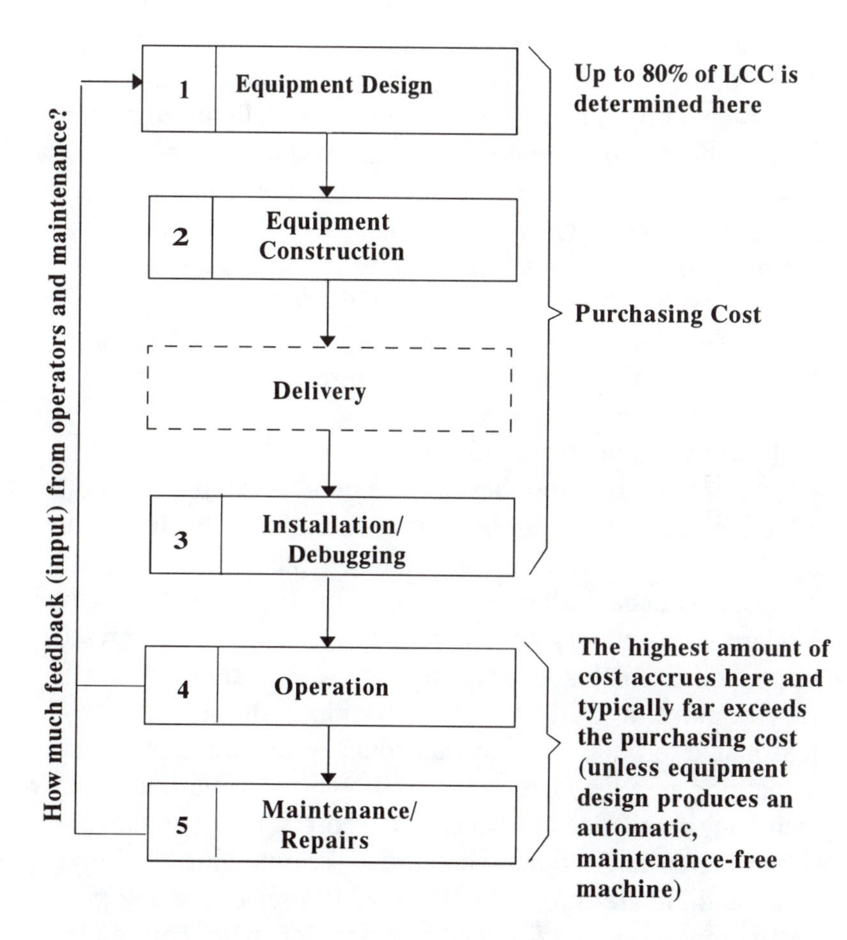

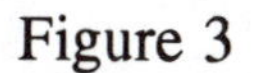
Figure 3

Have you ever considered how much input the representatives of the operations and maintenance departments, where most of the LCC accrues, have on the equipment design or specification function? The people who know most about the machines, have typically no -- or very little--input into the design or specification of their next machine! There is little chance to design out past problems if you don't know what those problems are and what impact they have on productivity and quality. It is difficult to design in any improved technology, because you may not know what changes will produce a better machine. The operators do, and maintenance does, because they work with these machines on a daily basis, and they know what works and what doesn't. They may not be able to tell you how to design these improvements into your machine, but they certainly can point you in the right direction.

Built-in diagnostics can lower LCC on all new equipment you procure. Automobiles are a good example of how effective diagnostics can be in preventing trouble. Indicator lights warn the driver of impending trouble from low oil pressure to a back door that is not properly closed. Many modern cars actually have a whole system of computerized inspections, which are completed before the driver takes off. If you have a million dollar machine, doesn't it seem reasonable to build in these watchdogs to let you know when something is going wrong? It is simple to do with existing technology and doesn't necessarily cost a lot of money. It doesn't even have to be computerized. Something as simple as an audible alarm or a light can alert your operators, who will know how to fix the problem, or who can notify maintenance.

Getting the Best From Your New Machines

Phase III of Equipment Management addresses the opportunities and the process that should be used to procure new equipment with a defined level of high performance and low LCC (Figure 4).

Step 1 is to develop the engineering specification (cycle time, level of automation, functions, etc.) of new equipment based on requirements of the new (or same) product. But steps 2 and 3 follow quickly, namely the input from operators and maintenance, based on their current equipment experience, supported by the equipment history. The next step is to eliminate past problems through better design or specifications, based on the knowledge gained in the previous two steps. The goal of user friendly (i.e. ergonomic) design must be kept in mind here. Don't forget quick changeover (or automatic changeover) capability to reduce or eliminate one of the equipment losses. New technology must be designed in during step 5, including the ones addressing safety and environment.

Step 6 is to design in diagnostics, such as oil pressure gauges, heat sensors, wear indicators, counters, fluid level sensors, vibration sensors, empty sensors, wrong location sensors, hour meters, trouble area indicators and counters, etc. A good example in the area of office equipment is copy machines, that not only display and indicate a problem, but pinpoint its location, display an error or problem code, and keep count of various types of problems (equipment history) for the service personnel to refer to. Some high-end machines even call for maintenance themselves after a certain number of problems!

EQUIPMENT MANAGEMENT

PHASE III

PROCURE (PURCHASE NEW EQUIPMENT WITH A DEFINED LEVEL OF HIGH PERFORMANCE AND LOW LIFE CYCLE COST

Step 1. Develop engineering specifications

Step 2. Get input from operators based on current equipment experience

Step 3. Get input from maintenance based on current equipment experience

Step 4. Eliminate past problems

Step 5. Design in new technology

Step 6. Design in diagnostics

Step 7. Design in maintainability (maintenance-free equipment)

Step 8. Start training (operational & maintenance) early

Step 9. Accept equipment only if it meets or exceeds specifications

Figure 4

Step 7 focuses on designing in maintainability, with the goal of maintenance free or at least maintenance friendly design. Examples of this process include easy access to cleaning and lubrication points, machine panels with clamps instead of screws, vacuuming equipment built into the machine etc. It is important to start the process of training on the new machine as early as possible (step 8). It is of great advantage, and highly motivational, to send maintenance personnel as well as operators to the equipment manufacturer for training, even before the machine is finished. Early and thorough training ensures that high standards of equipment performance and quality are established and maintained from the outset.

The final point (step 9) should not need any discussion, but experience shows that it does. Far too often, equipment is accepted from the manufacturer long before it meets agreed upon specifications. Often, the pressure from manufacturing to start using the new equipment is a contributing factor. Also, the time required for proper installation, debugging, and test run is often underestimated. The result is a new machine that starts out with a low OEE (Overall Equipment Effectiveness) and never (or only with great, delayed effort) comes up to the "defined level of high performance." Don't concede equipment performance, for which you have paid, for the sake of expediency.

TPM Goals

Part of improving and maintaining your equipment at its highest level of performance is to adopt ambitious goals. Like the "zero defects" goal of quality management, there are similar goals in TPM with regard to your equipment.

The Three Zeros

1. **Zero** unplanned equipment downtime

2. **Zero** (equipment caused) defects

3. **Zero** loss of equipment speed

The first and most difficult goal is zero unplanned equipment downtime. The reaction, when this comes up for discussion, is usually "impossible". However, the emphasis here is on *unplanned* downtime. How much planned downtime for *planned* maintenance, PM, cleaning, lubrication, inspection, and adjustment would you need to accomplish zero unplanned downtime?

Some Japanese automotive plants work eight hour shifts, then they are "down" for four hours, then another eight hour shift, etc. What happens during the four hours of planned downtime? Maintenance - planned maintenance! And cleaning, inspection, lubrication, etc. What happens to the equipment during the next eight-hour shift? Absolutely nothing; it just runs and runs. Zero unplanned downtime! This partially explains why Japanese automotive plants take much less time to produce a car than other plants.

Do you need 33% planned downtime to accomplish zero unplanned downtime? Certainly not. How much? You can establish that easily by following the steps of Phase II of equipment management (maintaining equipment at its highest level of performance) and adding the regular maintenance items to it. Operator involvement under TPM will reduce the time per day required for planned downtime.

Obviously, there is a point of "diminishing returns" and the accomplishment of "absolute zero" downtime may be cost prohibitive. But you must *believe* that zero unplanned downtime is possible and strive to accomplish it. If your maintenance management is data-based, you can establish where your break-even point is, and you will find that it is much closer to zero unplanned downtime than you think. Calculate the cost of unplanned downtime and compare it against the cost for additional planned maintenance to prevent it.

The second TPM goal is zero equipment-caused product defect. In some companies that strive for perfect quality levels, the equipment has become the barrier to accomplish this goal. The equipment must be in a condition that will not cause defects. *Perfect quality demands perfect equipment.* Companies that are serious about quality must also be serious about TPM.

Zero loss of equipment speed is the third goal. Speed loss is one of the "hidden losses", since the equipment speed (or cycle time) is usually not measured and compared to specifications. Sometimes theoretical speed or cycle time is not known and has to be established first. Very often, a speed loss is caused by worn out equipment that can't hold tolerances anymore at normal operating speed. The problem is, if this machine is part of a line, it slows down the whole line and throughput drops gradually over time. A 10% loss in machine speed is frequently found in industry today.

Those companies lose up to 10% in *productivity* that TPM can easily find and correct.

The Elements of TPEM

An approach used successfully in the non-Japanese world to install TPM is Total Productive Equipment Management (TPEM). The three components of TPEM are:

1. **TPM-AM** (focus is on autonomous maintenance)

2. **TPM-PM** (focus is on preventive and predictive maintenance)

3. **TPM-EM** (focus is on equipment management/ equipment improvement)

TPM-AM creates and organizes the operators' involvement in the care and maintenance of their equipment. Autonomous maintenance is the basis of the Japanese method, but plays a less dominant role in the Western world. It must be recognized that the cultural and managerial differences are significant enough that normally a different definition and approach to "autonomous maintenance" are developed outside Japan.

However, a degree of "autonomous maintenance," as part of your small group structure, is important and critical to TPM's success. TPM-AM lets you create the type and amount of operator involvement that fits your corporate and plant culture and responds to the needs of your equipment and organization (see Chapter VII for more details).

TPM-PM, including both preventive and predictive maintenance, is a total system of PM for the entire life of the equipment. There is a degree of overlap with TPM-AM, as operators are expected to participate in the preventive maintenance of their equipment, eventually executing Type I PM in an autonomous fashion. However, a good PM system must be developed and executed under Total Productive Equipment Management, regardless of the degree of operator participation (TPM-PM is covered in detail in Chapter VIII).

TPM-EM is a very successful approach to quickly improve your equipment's performance and to get your operators initially involved in TPM. It is a very profitable, exciting and often fun part of TPM. TPM-EM is usually the first component installed in a plant, where improvement of equipment performance is a high priority. It will give you a good indication of the talent and potential you have in your operators and maintenance personnel, forecasting their ability to successfully complete the total TPM installations (Details of TPM-EM are covered in Chapter IX).

The Meaning of Total

What exactly does the "Total" in TPM mean? First, it stands for total economic effectiveness. That means you are looking for profitability at a level you have never before achieved. TPM more than pays for itself. ROI (Return on Investment) levels of 400% and more have been accomplished outside Japan.

TPM also includes total coverage. You are going to

examine and address all your equipment, not just the bottlenecks, and strive for the best possible maintenance and equipment performance on a plant-wide basis. Obviously, the sequence of your installation will concentrate on and improve your "production limiting" equipment first.

"Total" encompasses all aspects of a maintenance system. Not just preventive or predictive maintenance, but planned maintenance, computerized maintenance (CMMS), maintenance controls, good planning and scheduling, and all other maintenance techniques that are at your disposal.

"Total" means full participation by all involved employees. TPM is not a system for the maintenance department or just one shift. To work properly, all involved employees must know and practice TPM to the best of their ability. It specifically includes management, who must provide leadership, guidance and support, recognition and a system of rewards.

In later phases, "Total" means not just your equipment and production areas, but other departments involved with equipment; such as engineering, purchasing, etc. In some plants, TPM is extended to cover general offices.

The TPM Organization

To successfully install TPM in your plant or corporation, it must be supported by an organizational structure that will enable its progress and success. Very often, the need for this organization is not recognized (or too late), slowing down the progress of an otherwise good TPM installation.

Figure 5 shows a typical TPM organization in schematic form. *The line organization* reaches from top management, which participates in the corporate or plant steering committee, down to the TPM small groups on the production floor.

The *line organization* includes a plant TPM champion, usually a high level manager, who will have overall TPM responsibility. However, the most crucial person to your TPM success is the TPM Manager, sometimes called the TPM Coordinator. This job is typically a *full time function*,

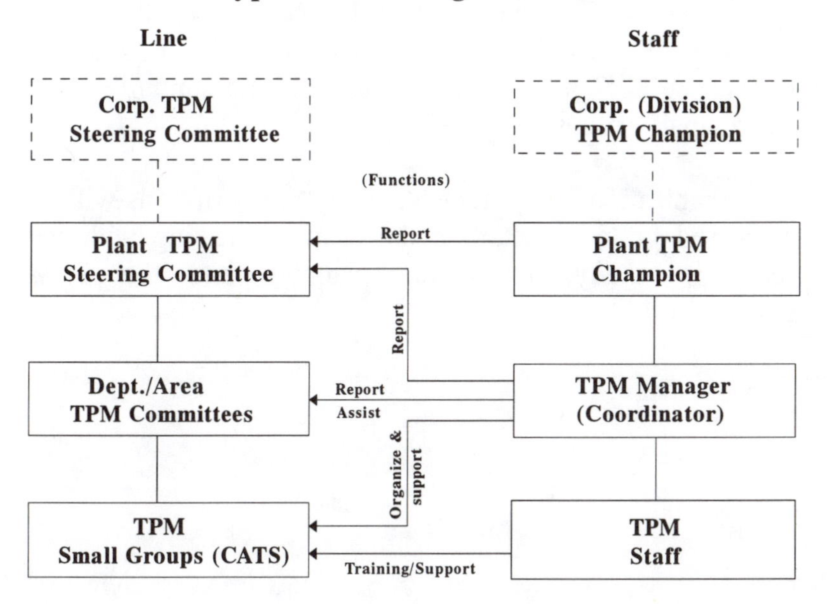

Figure 5

except in a small plant, that has the responsibility for planning and executing your TPM installation. The TPM manager develops and often conducts the TPM training, guides the execution of the feasibility study, measures and reports progress, and promotes TPM in your plant. Additional jobs are to support the various TPM small groups and to provide liaison between these groups, between maintenance and production and between the TPM organizations and management. Truly, a most important and critical function, whose proper staffing is often the key to a company's TPM success.

Depending on the size of your plant and the workload of your TPM manager, there may be a need for a full or part-time TPM staff, supporting and reporting to the TPM Manager. The staff is often involved with developing and delivering TPM and skills training and with supporting the TPM small groups.

Organizational requirements and job descriptions will be discussed in more detail in the chapters dealing with the feasibility study, installation planning and the TPM installation.

CHAPTER IV

The Power of TPM

Although TPM is in its relative infancy, there are already a number of success stories to tell. The Japanese, of course, have been practicing TPM for about 20 years. Today, it is estimated that over 1000 Japanese plants use TPM, covering the whole spectrum of industry, from micro electronics to automotive and steel making.

While the Japanese started TPM, they have no corner on the market. The trend to TPM is international. There is a tremendous surge of interest in TPM in Latin America, Southeast Asia and Europe.

In the U.S. too, TPM is catching on at a number of large companies. Ford Motor Company, Eastman Kodak, DuPont and Motorola are some of the leading corporations that are now installing TPM programs in many plants both in the U.S. and overseas.

TPM Impacts all of Manufacturing

Most of the results are outstanding. And they occur in all phases of the manufacturing process. In one U.S. aerospace company, implementing TPM helped them reduce maintenance service calls by 29% in only three months.

Of course, the primary purpose of TPM is to reduce equipment downtime. The reason is simple. You only make money when your equipment is running. Waiting for maintenance and fixing breakdowns is costing you precious production time. So you must prevent breakdowns and eliminate unnecessary idling and stoppages of equipment. You have to train and motivate your operators to participate in accomplishing these goals.

Just these four reductions--fewer equipment failures, quicker changeovers, less maintenance downtime and less idling and minor stoppages--can give you 40% more output in the same time. That's like picking up 24 minutes of extra production time for every hour your machines are operating.

Using TPM, you can increase equipment speed by about 10%. One of the major reasons for equipment slowdown is worn parts. Another cause of speed loss is loose bolts or screws on the machine. Vibration causes these fasteners to work loose. There is vibration on any machine that has a motor or other rotating and oscillating parts. Some of that can be cut down by balancing bearings, gearboxes and shafts. But even the newer high precision machines have vibration. So tightening bolts and screws is one routine chore that will pay big dividends in machine speed.

Lubrication is the lifeblood of equipment operation and speed, yet it's often neglected. Operators can inspect their machines, maintaining a checklist to ensure that routine maintenance is done on a regular basis. All of these actions keep the equipment in better condition so it can be run at higher speeds.

Reducing Defects

TPM has cut the defect rate by 90%, from ten per thousand to one per thousand at the Tochigi plant of Nissan. Your quality can increase from 99% to 99.99%. Some Ford and Motorola operations are doing it already. That's very close to the zero defects goal.

Regular maintenance is the key, and record keeping is how you ensure that PM and other maintenance is performed on schedule. Many quality-conscious companies are already using Statistical Process Control (SPC). Operators trained in SPC do statistics, plot charts, and perform other paperwork. Years ago, if you had asked operators to do this job, they would have said it was impossible. Today conditions are different.

Most operators who are properly motivated will also inspect their equipment on a regular basis. Again, you need training to produce this response, but once operators become involved with their equipment, they will want to inspect it to ensure that it's in good condition.

The bottom line in this quality process is equipment improvement and uncompromising maintenance. By making sure that your equipment is in top operating condition, you have a much better chance of producing a quality product. And that's what it takes to be competitive in today's global marketplace.

The Passion of Productivity

Improved equipment quality and performance lead to improved productivity. Dai Nippon in Osaka, Japan, has accomplished a *plant-wide* productivity gain of 50%. These benefits came through fewer breakdowns, less idling and minor stoppages, shorter set-ups, faster speed and fewer rejects.

Suppose you are currently producing 1000 parts or components per day, and you could increase that to 1500, *without* adding an extra shift. What effect would it have on your company's earnings? That's the power of TPM.

Normally, you can't expect your whole plant to attain that 50% improvement. But on many machines you can. Establish a goal that you can reach by studying the current conditions in your plant and calculating the overall effectiveness of your equipment, then determining how much you can improve it and what your new output will be.

Controlling Maintenance Costs

Robotics, automated factories, computer-integrated manufacturing, computerized numerical control (CNC)--all these high-tech accomplishments are helping companies produce more and better quality products. But these new, complicated machines that are part of this technology are expensive to buy, repair and maintain. So the demands on maintenance and maintenance costs are soaring wherever this new technology is installed.

TPM can help you control maintenance costs. Cost reductions of 30% for plants where TPM is installed have been reported. Sometimes you can get that 30% in one area alone, maintenance travel and delays. The operator is already there, and with proper training, can fix many problems, eliminating a large portion of travel time.

Delays can eat up 35% of a maintenance worker's productive time. You schedule some maintenance work on a machine, the maintenance craftsman gets a job ticket and goes to the site. However, there's a production run that can't be interrupted. The maintenance worker waits--and waits. You're paying this highly-trained and highly-paid expert to sit around and watch a production line run. If this job could be done by the operator, it could be scheduled conveniently during a production break with no time lost.

Figure 6 illustrates how you turn a maintenance department into a high tech operation, using TPM. Delegate the routine work, such as equipment cleaning, adjusting, lubricating and set-up, to the machine operators. You can even turn over many inspection tasks, some or most of your preventive maintenance, and possibly a few minor repair tasks.

That frees up the maintenance worker to invest more time in high tech activities such as equipment monitoring and improvement. Qualified craftsmen should do more major PMs and needed equipment overhauls or rebuilds, for which there never seems to be sufficient time available. Predictive maintenance, to determine equipment condition and needed repairs, is another high tech job for these specialists. Even assisting in new equipment design is within the scope of these craftsmen.

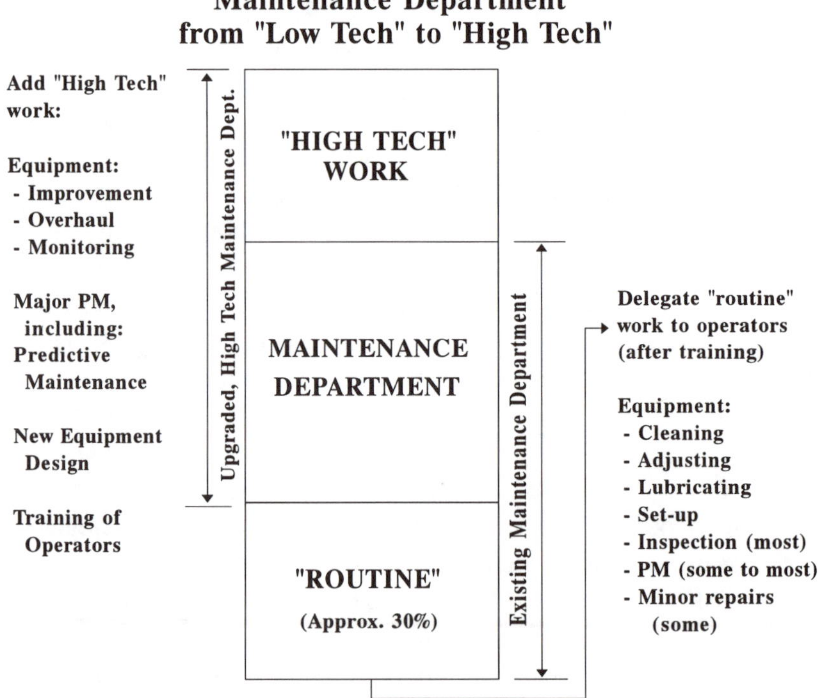

Figure 6

Part of this new, high-tech maintenance operation is the training of the operators, which becomes important under TPM. When maintenance workers realize the benefits of transferring their routine work to operators, training will receive high priority.

Improving Your Safety Record

Another benefit of TPM is increased safety. In addition to zero defects, the goal of TPM is zero accidents. Tennessee Eastman, a chemical company that has the first and most successful TPM installation in the U.S., suffered only three minor accidents while performing over 1,000,000 TPM tasks (a task that was previously done by maintenance) over the last four years. This is a vast improvement over their previous record. Under TPM, the operators are trained and motivated to work safely. If one operator is unsure of how to perform a TPM task correctly, another, more experienced operator will pitch in and help. That's the team concept and the reason safety improves dramatically with TPM.

The Bottom Line

The return on investment normally pays for your TPM program many times over. Dai Nippon, a large Japanese printing company, invested $2.1 million in TPM. But the company saved $5.5 million over the same period, an ROI of 262%. Tennessee Eastman spends $1 million annually on TPM. Their documented *cost reduction* is over $5 million a year, an ROI of over 500%. This does not include the benefits of improved productivity (output), which are estimated to be a multiple of the cost reduction benefits!

You have to make some investment in TPM to make it work. It is no quick cure by any means. There are costs for TPM administration, for training, and for equipment improvement. You must be able to calculate where your savings are going to happen, and approximately how much you can expect to gain.

Here's an example from the Tennessee Eastman plant to give you an idea of how to calculate what you save. Some of their equipment includes a small rubberized disk that serves as a safety valve. If the vacuum in the machine (a chemical reactor) gets too high, the disk ruptures and the machine shuts down. Before TPM, it took four hours to get the equipment back on line again. Why so long? The machine operator notified the production supervisor, who notified the maintenance supervisor, who pulled someone off a current job to fix the problem. The craftsman then went to stores to get the replacement part, took it to the machine, then removed four bolts on two flanges. He took out the ruptured part, put the new one in, aligned it, and refastened the bolts. This happened over 200 times a year. That translates into 800 hours spent replacing a safety valve at an annual cost of about $20,000.

When TPM was installed, the operators decided (along with maintenance) that this was a task they could do themselves if they had the right tools and parts available. So, after a period of training, they took on the job. And a strange thing happened. The number of the safety disk failures fell to 20 a year and a year later to 10. The operators didn't feel like fixing this disk 200 times a year. So they began watching their dials much closer to prevent the vacuum from building too high and ripping the disk. And because the operator had the tools and the replacement part at the job site, equipment downtime dropped from four hours to one.

The new calculation (10 times one hour per occurrence times $25 per hour) yields a total cost of $500 for repairs. That's a cost reduction of $19,750 or 99%. It also gives the maintenance department 800 more hours to devote to other work. And production adds 790 hours to its uptime, which alone is quite significant and is not even included in the savings calculation. Multiply that cost reduction figure by hundreds of other maintenance tasks that operators do in the plant, and the ROI becomes truly impressive.

Employee Participation

One overriding result of TPM is employee pride in performance. In every mature TPM installation, the operators are proud of their accomplishments. They'll walk up to you and tell you how they've improved their machine. It's really an intangible in terms of cost reduction or improved performance, but it's there and you can see it.

Your employees will have more job satisfaction because they are involved with the equipment. They'll develop a sense of "ownership" of a machine, which contributes to this sense of pride. Teamwork will also increase. You'll have more interaction, more brainpower brought to bear on solving maintenance and equipment problems. And team members will support each other on the job.

Because TPM requires training, you'll find your employees will have improved skills. With the rapid explosion of technology, enhanced skills are going to be a big plus in the next ten years. For many companies, the quantum leap in automated equipment is going to require employees who are versatile and better trained.

This employee involvement in the TPM process results in less turnover. Your work place will be more interesting because the employees have a stake in the work process. Your greatest asset, people, will be working with you to improve quality and productivity, and reduce equipment failure and lost work time.

Putting the Power of TPM to Work

TPM works extremely well in getting the employee involved with the process. But you can't expect this involvement and motivation to happen automatically. Money, of course, is a prime motivator in the U.S. and many other countries. The more skills workers have, the more pay they can receive. And even though you may pay workers at a higher classification, the cost to accomplish increased skills is a good investment. Of course, there are non-monetary incentives. It may be in the form of awards or a luncheon when certain goals or milestones are reached.

What you always have to keep in mind is that TPM takes time, commitment, training and motivation. There likely is going to be a certain amount of resistance to change, sometimes more so if a union or unions are involved. This reaction is normal and found everywhere in the world.

An electronics company in Malaysia used an interesting approach of explaining TPM to their operators, who were mostly young women. They compared TPM to a mother caring for her baby. The equipment, of course, is the baby. The mother, the operator, was responsible for keeping the baby clean, feeding it, lubricating it, etc. The mother also monitors the baby. When it cries, something is wrong. The mother must take its temperature to see if the baby is running a fever.

The mother does a lot of TPM activities, but what if she runs out of options? Then it is time to call the doctor. The doctors are the maintenance people, the specialists. They go in and fix the baby (the machine) if the mother can't make it well again.

Companies can use TPM to make their operations healthy again. The healing power is there. But like any medicine that cures an illness, it must be used according to the prescription. Not enough dosage and the patient does not improve. Too much at a time and the user will have a bad reaction that wipes out all the good that was intended.

That's why careful analysis of your current situation (the feasibility study), the custom-made design of your TPM program and a well-organized and managed installation are so important.

CHAPTER V

Measuring Your True Equipment Productivity

Unlocking The Hidden Factory

Within most plants around the world there lies a hidden factory. Occasionally you catch a glimpse of it, when production is humming along and everything is going right and no machine is down. You know it's there, just below the surface, the potential of what your plant could be if everything would just continue to work as it should. You wish it could be that way all the time, but somehow problems get in the way and it vanishes, screened from your sight by the reality of everyday business.

TPM is the key that can unlock that hidden factory and bring perhaps another 25 to 30% of capacity into your production areas. Here is how you calculate your current equipment productivity and determine your improvement potential.

Equipment Productivity

True equipment productivity is measured by Total Effective Equipment Productivity (TEEP).

This is the overall formula that includes Equipment Utilization (EU) and Overall Equipment Effectiveness (OEE). Most of the current TPM literature discusses only OEE and disregards the fact that a high level of equipment utilization is required to accomplish a high degree of equipment productivity and a good Return on Assets (ROA). You can improve your OEE at the expense of equipment utilization by doing all your set-ups and PMs during planned downtime. If plant management is truly interested in getting good asset and capacity utilization, the TEEP formula is of prime importance (Figure 7).

Total Effective Equipment Productivity (TEEP), with the emphasis on "effective productivity", includes planned downtime and is a combined measure of equipment utilization and overall equipment effectiveness.

Overall Equipment Effectiveness (OEE) is the traditional and most widely used TPM measure. It reflects how the equipment is performing overall *while it is being operated.* As a matter of fact, it is not an *exact* measure of the equipment effectiveness, since set-ups or changeovers and resulting adjustments are included. This does not have much to do with the equipment's performance itself, but reflects the overall equipment's effectiveness while the equipment is being run.

The Three Major TPM Formulas

TEEP (Total Effective Equipment Productivity)

= Equipment Utilization (EU) **x**
Overall Equipment Effectiveness (OEE)

OEE (Overall Equipment Effectiveness)

= Equipment Availability (EA) **x**
Performance Efficiency (PE) **x**
Rate of Quality (RQ)

NEE (Net Equipment Effectiveness)

= Uptime (UT) **x**
Performance Efficiency (PE) **x**
Rate of Quality (RQ)

Figure 7

Therefore a third formula, that clearly reflects the *true quality and effectiveness of equipment while it is running*, seems to be in order. Net Equipment Effectiveness (NEE) is this formula.

It excludes not only planned downtime (as does OEE), but also downtime required for set-ups and adjustments. It is a reflection of the true mechanical condition of your machine.

Equipment Losses

In order to calculate these three indices--TEEP, OEE, and NEE--you need to know what your equipment losses are. TPM focuses on equipment losses that cut into your equipment effectiveness. There are at least five categories:

- Set-up and adjustments
- Equipment failures
- Idling and minor stoppages
- Reduced speed
- Process defects (see Figure 8)

In many companies, there are more, such as warm-up losses, test runs, etc. Those losses must be identified beforehand and included in the appropriate formula. It has been found that the "reduced yield" or "start-up" loss (the difference from equipment start-up to stable production) as described in other publications can not be measured as such, since it normally consists of a combination of above five losses during the equipment de-bugging or start-up phase. It is recommended to calculate the OEE at equipment

The EQUIPMENT LOSSES
you can and must measure

Equipment Availability	**Set-up and Adjustments** - Including changeovers - Programming - Test runs **Equipment Failures** - Sporadic breakdowns - Chronic breakdowns
Equipment Efficiency	**Idling and Minor Stoppages** - Jams and other short stoppages - No parts, no operator - "Blocked" - Many other reasons **Reduced Speed** - Equipment worn out - Lack of accuracy
Quality	**Process Defects** - Scrap - Rework
	Others (define) - Equipment warm-up, etc.

Figure 8

installation and then again at stable production to determine the "yield loss".

In the semiconductor industry, the term "yield" is used for the percentage of usable chips obtained from a wafer and could be used as part of the OEE formula under Quality Level. However, caution is advised, since this has no connection with the actual effectiveness of the machine in question.

The first equipment loss is set-up and adjustment. When you do a set-up, the machine is down, although it is not broken down. Of course, it's a necessary part of production, but since it is a *variable* and can be reduced, it does qualify as an equipment loss. Frequently, set-ups and changeovers are among the largest equipment losses, indicating the need to carefully measure this loss and to develop improvements.

Unplanned downtime (equipment failures) is next. There are two types of equipment failures: sporadic and chronic. Sporadic failures happen suddenly. Something on the machine breaks. Usually you can identify it easily and fix it. It normally doesn't re-occur often. Chronic failure is more difficult to deal with. Every once in a while, the machine stops and you may not even know why. You suspect the cause, but you can't pin it down. Eventually the plant learns to live with the defect. This compromise is not the right solution and is not allowed to happen under TPM.

Both of these first two losses figure in the measurement of *equipment availability*. In each case, the machine is down and therefore not available for production.

The next two losses are also called "hidden losses." They're usually not measured and not recorded as downtime because maintenance is not called and the equipment is not broken down. It just runs less efficiently.

Idling and minor stoppages falls into this category. The machine's motor is running, but no product is being processed. Perhaps there is a jam and no product is coming into the machine, or the machine next in line is down and you are "blocked," or the operator is not available for a few moments. Maybe you are momentarily out of parts, or the machine is out of adjustment and needs to be re-adjusted. There are so many reasons for idling and minor stoppages.

These little problems can cause some of the biggest losses in a factory. In one electronics plant in Asia, a female operator was testing electronic parts that came down into the machine through a channel. Every so often, the machine stopped (jammed) and the operator used a small tool like an oversized toothpick to get it running again. It only took her about four seconds to fix the problem, which happened on the average of *three times a minute*. That's 12 seconds, and if you stop to figure it out, it's 20% of each production minute. Multiply those 12 seconds per minute by eight hours and you have a considerable loss of production. Jams figure prominently on every chart of idling and minor stoppages analysis and frequently account for a high percentage of loss. Yet the reasons for most jams are relatively easily corrected.

Reduced speed is the fourth major equipment loss. It stems mainly from poorly-maintained, worn out or dirty equipment. Some other causes of speed losses are insufficient debugging during the start-up phase, defective mechanisms or systems, design weaknesses and insufficient equipment precision.

These two losses figure in the calculation of *performance efficiency*. In each case, the machine is not broken down, but performs at a lower level of efficiency.

The fifth equipment loss is process defects. If a part is rejected or must be reworked, the equipment time producing it is lost. This loss is relatively small when compared to other major equipment losses. However, in today's environment of Total Quality, no rejects, especially those caused by a machine, are tolerable. Typically, as the equipment gets improved and better maintained under TPM, quality losses are also reduced. Nevertheless, the reason for each quality loss must be investigated and the equipment problem causing it must be eliminated. This loss is used to calculate the *rate of quality*.

As discussed before, there may be other losses in your plant. You must identify these during the feasibility study and include them in your calculations.

Measuring each of these losses will determine overall equipment effectiveness (OEE) and net equipment effectiveness (NEE) of your machines. Without proper identification and quantification of your equipment losses, it will be very difficult to establish an effective and tailor-made TPM program.

Calculating Equipment Effectiveness

Once all these losses are known, you can calculate your equipment effectiveness on a step-by-step basis. Figures 9 and 10 show the procedure and a typical example.

Equipment is sitting in your plant 24 hours a day. Therefore, start with the total available minutes (1440) in a 24-hour day. The company used as a typical example here is running two shifts, so subtract 480 minutes (8 hours) for one shift. Then subtract the planned downtime, which includes breaks and meals for the other two shifts plus planned maintenance and any other planned downtime, such as meetings and no scheduled production. The calculation will establish your percentage *equipment utilization* (60.4%).

The remaining time left after deduction of unutilized time is called running time (870 minutes). At this point, the OEE calculation starts, since now actual equipment losses come into play. First, deduct the time spent for set-ups, changeovers and adjustments (70 minutes). The resulting calculation will give the *planned availability* (92.0%), which is one part of the equipment availability (EA).

The time left after deduction of the set-ups is the operating time. At this point, the calculation of the Net Equipment Effectiveness (NEE) starts. The amount of time the equipment was broken down due to failures (unplanned downtime) is now deducted and the percent *uptime* (93.7%) can be calculated. Unfortunately, this is often the *only* number reported to plant management, creating a *totally wrong* impression of the real equipment situation, since it only covers one single loss. For that reason, plant and

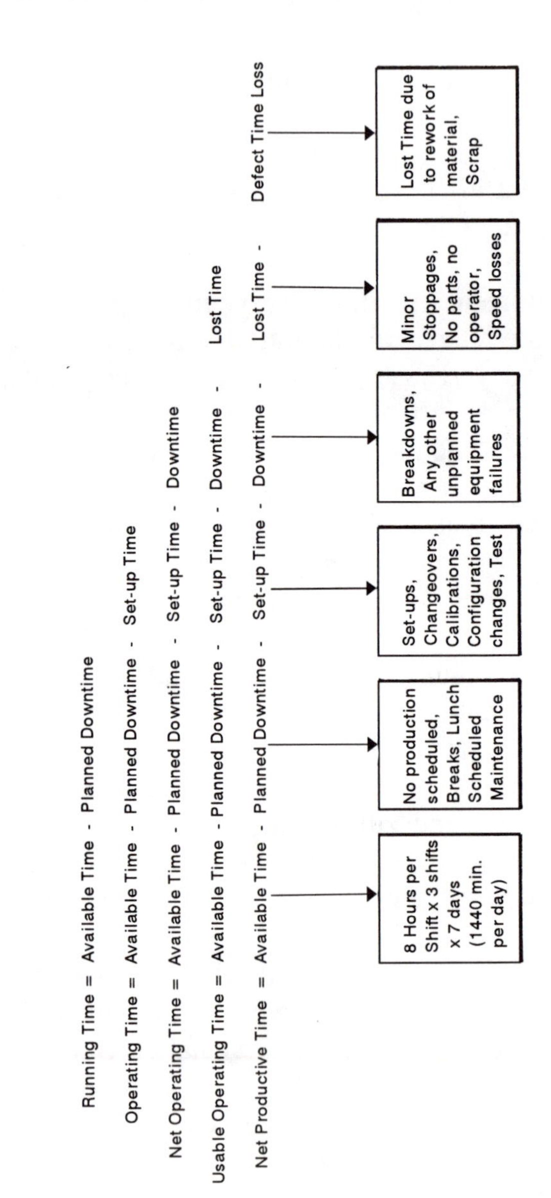
Definitions for Equipment Loss Calculations
Running Time = Available Time - Planned Downtime
Operating Time = Available Time - Planned Downtime - Set-up Time
Net Operating Time = Available Time - Planned Downtime - Set-up Time - Downtime
Usable Operating Time = Available Time - Planned Downtime - Set-up Time - Downtime - Lost Time
Net Productive Time = Available Time - Planned Downtime - Set-up Time - Downtime - Lost Time - Defect Time Loss
8 Hours per Shift x 3 shifts x 7 days (1440 min. per day)
No production scheduled, Breaks, Lunch Scheduled Maintenance
Set-ups, Changeovers, Calibrations, Configuration changes, Test
Breakdowns, Any other unplanned equipment failures
Minor Stoppages, No parts, no operator, Speed losses
Lost Time due to rework of material, Scrap

Figure 9

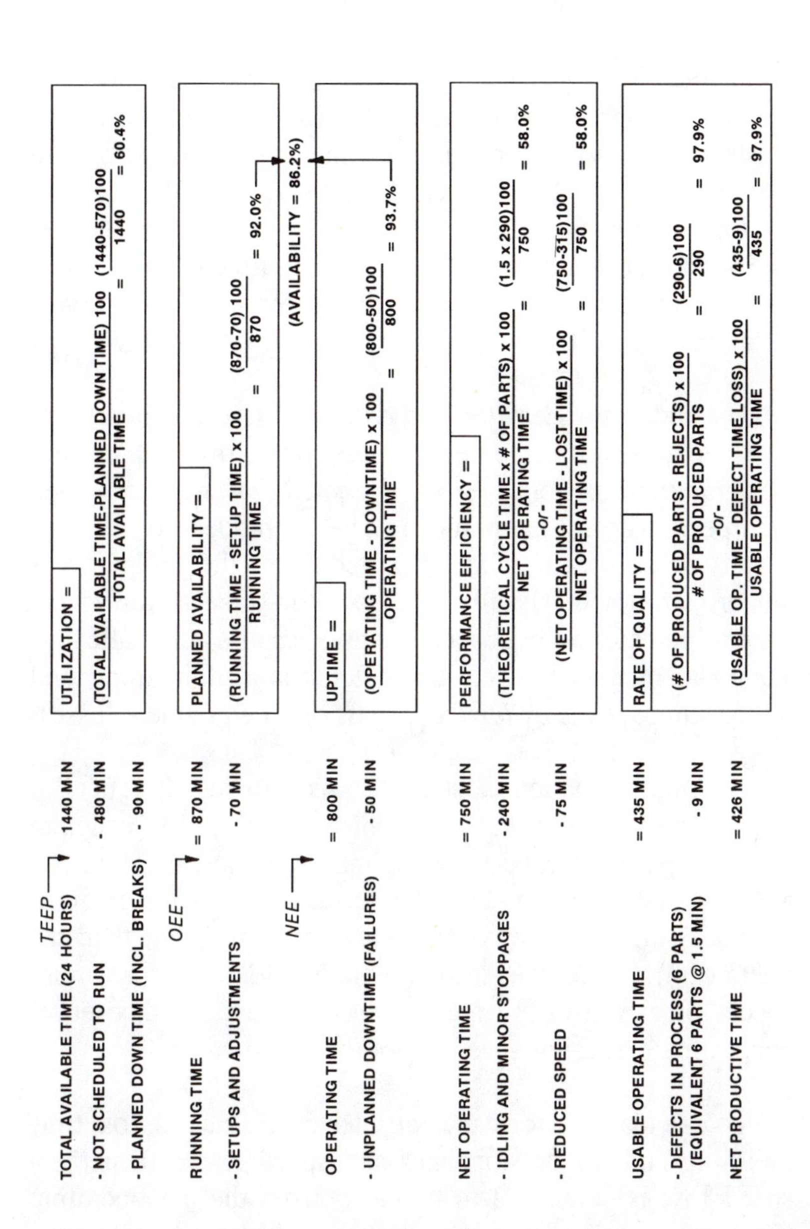
TOTAL AVAILABLE TIME (24 HOURS)
TEEP
1440 MIN
- NOT SCHEDULED TO RUN
- 480 MIN
- PLANNED DOWN TIME (INCL. BREAKS)
- 90 MIN
OEE
RUNNING TIME
= 870 MIN
- SETUPS AND ADJUSTMENTS
- 70 MIN
NEE
OPERATING TIME
= 800 MIN
- UNPLANNED DOWNTIME (FAILURES)
- 50 MIN
NET OPERATING TIME
= 750 MIN
- IDLING AND MINOR STOPPAGES
- 240 MIN
- REDUCED SPEED
- 75 MIN
USABLE OPERATING TIME
= 435 MIN
- DEFECTS IN PROCESS (6 PARTS)
(EQUIVALENT 6 PARTS @ 1.5 MIN)
- 9 MIN
NET PRODUCTIVE TIME
= 426 MIN
UTILIZATION =
(TOTAL AVAILABLE TIME-PLANNED DOWN TIME) 100 / TOTAL AVAILABLE TIME = (1440-570)100 / 1440 = 60.4%
PLANNED AVAILABILITY =
(RUNNING TIME - SETUP TIME) x 100 / RUNNING TIME = (870-70) 100 / 870 = 92.0%
(AVAILABILITY = 86.2%)
UPTIME =
(OPERATING TIME - DOWNTIME) x 100 / OPERATING TIME = (800-50)100 / 800 = 93.7%
PERFORMANCE EFFICIENCY =
(THEORETICAL CYCLE TIME x # OF PARTS) x 100 / NET OPERATING TIME = (1.5 x 290)100 / 750 = 58.0%
-or-
(NET OPERATING TIME - LOST TIME) x 100 / NET OPERATING TIME = (750-315)100 / 750 = 58.0%
RATE OF QUALITY =
(# OF PRODUCED PARTS - REJECTS) x 100 / # OF PRODUCED PARTS = (290-6)100 / 290 = 97.9%
-or-
(USABLE OP. TIME - DEFECT TIME LOSS) x 100 / USABLE OPERATING TIME = (435-9)100 / 435 = 97.9%

Figure 10

production managers are usually perplexed when they hear the real equipment effectiveness (OEE) numbers after the feasibility study. Uptime is the other part that makes up equipment availability. *Availability* is determined by multiplying planned availability (92.0%) by uptime (93.7%) = 86.2%. Or you can divide the running time (870 minutes) into the remaining net operating time (750 minutes) to arrive at the same result.

The index for *performance efficiency* is calculated next. The starting point is the net operating time, from which first idling and minor stoppages are deducted (240 minutes), then the speed losses (75 minutes). These "hidden losses" are usually never measured and reported, since the equipment is not broken down. Usually, the operators take action to get the equipment running, or the machine restarts automatically. To make matters worse, it is often found that idling and minor stoppages is *by far the largest* of all equipment losses!

A similar situation exists with speed losses. Frequently, the speed of a deteriorating machine is reduced in order to maintain parts tolerances, or the machine just cannot be run at full speed anymore. Usually, those speed losses creep in gradually and nobody is acutely aware of it (other than the operators), hence the second "hidden loss." In addition, speed losses are rarely measured and often, the theoretical cycle time or design speed is not even known.

To calculate performance efficiency, deduct the lost time of idling and minor stoppages and speed losses (a percent speed loss is converted to minutes) from the net operating time and then compare the resulting usable operating time with the net operating time (58.0%).

Another formula (used by Nakajima) is the theoretical cycle time multiplied by the number of parts produced over net operating time. However, it was found that this formula is often difficult to use. Sometimes the theoretical cycle time is not known, or different products with different cycle times are run on the same machine, making it hard to use this formula. Using minutes of lost time is much simpler during the observation period and for the calculation.

The last calculation determines the *rate of quality*. The defect time loss (number of defective or reworked parts x time per part) is deducted from the usable operating time, resulting in the net productive time. This number is then compared to the usable operating time to establish the rate of quality (97.9%). The advantage of this procedure is that only a single unit of measure (minutes) is used throughout all calculations, making it simple to program the computer to perform all calculations. The other formula uses the number of rejects and the resulting net amount of good parts is then compared to the total number of parts produced for the same results.

Where do those numbers come from? A team of observers must collect them as part of your feasibility study, the first step before installing TPM in your plant. Use of mechanical or computerized measurements of equipment for this study is not practical, since it is very difficult to distinguish the exact losses. The observers need to concentrate on such items as set-ups and adjustments, equipment failures, idling and minor stoppages. The rate of quality is usually calculated comparing the number of rejects to the total number of produced parts. Speed losses are often expressed as a percent of optimum speed.

From this monitoring, a *baseline* of your current equipment effectiveness and other data will be developed. This analysis will also point toward the areas where the biggest problems, and therefore biggest opportunities are. This allows you to put your efforts into improvement activities that will give you the greatest operating benefits.

Applying The Formulas

Now you can apply these numbers to the three equipment effectiveness formulas (see Figure 11). Plant management should set their TPM goals based on these results.

To determine the Total Effective Equipment Productivity (TEEP) in this example, multiply equipment utilization (60.4%) by equipment availability (86.2%) by performance efficiency (58.0%) and by the rate of quality (97.9%). The result is only 29.6%! The same result is determined by dividing the total available time of 1440 minutes into the net productive time of 426 minutes (the net amount of equivalent time the machine is *actually* producing good parts).

OEE is the same calculation, but excluding equipment utilization, and the result is 49.0%. This number represents the effectiveness of the equipment while it is being operated. The same number can also be obtained by dividing the running time (870 minutes) into the net productive time.

Net equipment effectiveness (NEE) excludes the set-ups from the OEE. Multiply uptime (93.7%) times performance efficiency, times rate of quality. This figure reflects the true quality of your equipment (53.2%).

Equipment Productivity and Performance Calculations

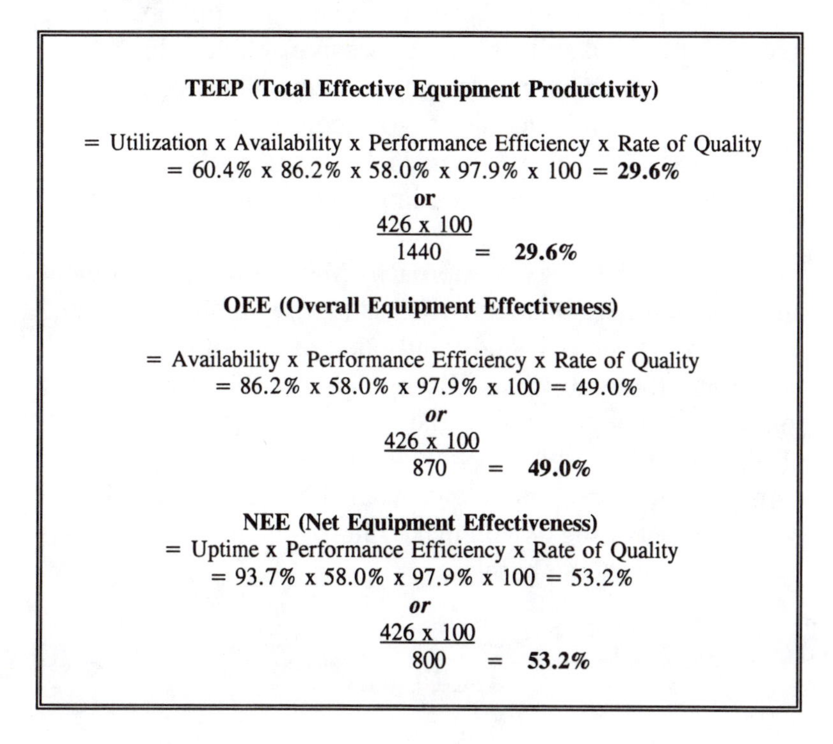

TEEP (Total Effective Equipment Productivity)

= Utilization x Availability x Performance Efficiency x Rate of Quality
= 60.4% x 86.2% x 58.0% x 97.9% x 100 = **29.6%**

or

$$\frac{426 \times 100}{1440} = \mathbf{29.6\%}$$

OEE (Overall Equipment Effectiveness)

= Availability x Performance Efficiency x Rate of Quality
= 86.2% x 58.0% x 97.9% x 100 = 49.0%

or

$$\frac{426 \times 100}{870} = \mathbf{49.0\%}$$

NEE (Net Equipment Effectiveness)
= Uptime x Performance Efficiency x Rate of Quality
= 93.7% x 58.0% x 97.9% x 100 = 53.2%

or

$$\frac{426 \times 100}{800} = \mathbf{53.2\%}$$

Figure 11

Unfortunately, the numbers in this example are quite representative of most equipment today, even in relatively new plants! In most cases, you are not getting anywhere near what your equipment should give you.

Based on the numbers for your plant, you now have choices. Many companies schedule a third shift to get more output. That increases the utilization number in the formula from 60% to 90%. But there may be another option. Hiring people for a third shift is very expensive and you don't have planned downtime anymore to do your maintenance. Instead, you can focus on bringing up your equipment effectiveness (OEE) from 49% to perhaps 75% on two shifts to get the same results (a 50% increase in output) at much less cost!

TPM is a data driven approach. Manage your equipment and your improvement activities using the numbers obtained during the feasibility study and later re-measurements. It's important to establish a baseline to be able to see your improvement opportunities. Using the three formulas--TEEP, OEE and NEE--you get the numbers that allow you to manage your equipment and your business. You should continue measuring as you make improvements to chart your progress against the baseline.

OEE Goals

What should a good TPM program accomplish? Many "world class" companies reach 85% overall equipment effectiveness after a successful TPM installation (Figure 12). Equipment *availability* should be more than 90%. Using the example presented, attaining this number requires only about 4% over the current number. The *rate of quality* should go up to at least 99%, about 1% above the example. *Performance efficiency*, however, must jump from 58% to more than 95%. Obviously, here is the big improvement opportunity. And what drags down performance efficiency? Idling, minor stoppages and the speed loss of the equipment (the "hidden" losses!). Typically, equipment inefficiency lies

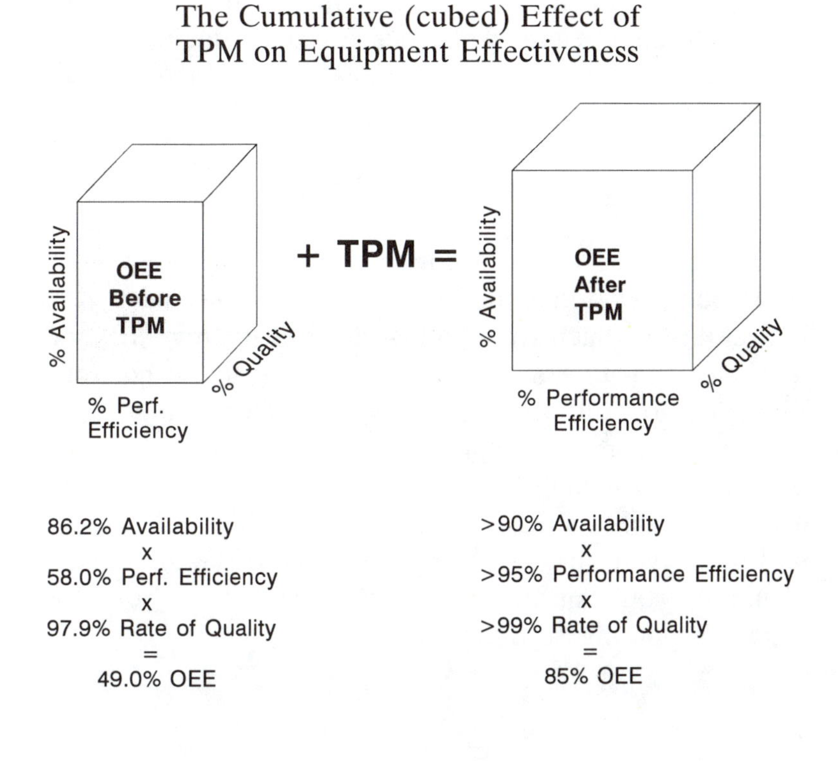

Figure 12

in these areas, and you will get many unexpected surprises when you start to investigate these factors. But this is the area where the greatest potential lies for productivity gains.

A new OEE of 85% represents a productivity improvement of 73.5%, using the example. It's not likely that a figure this high can be accomplished plant-wide. It is achievable with some individual machines. However, a 50% plant-wide productivity improvement has been reported by successful TPM companies, accomplished primarily through improvements in equipment efficiency and reduction in set-up losses.

Setting Your Priorities

The numbers obtained during the feasibility study will help you set priorities for TPM. Very few companies have the resources which would allow them to improve *all* their machines and *all* their production lines at the same time. There are normally limitations of manpower and money, so it is quite important to apply them wisely and to set the *right priorities* to accomplish a quick return on investment (ROI) and measurable increase in *throughput*. These calculations will help you make the right decisions. With careful planning, you can get maximum productivity from the resources you invest in the program.

CHAPTER VI

Customizing Your TPM Installation

TPM Strategy

TPM is much more than a maintenance program. It is an equipment improvement and management program. This distinction makes quite a bit of difference in how you approach the installation of TPM.

Total Productive Equipment Management (TPEM), the International TPM Institute's process of TPM installation, is more readily accepted by employees if you introduce it as equipment management. If you say you're going to install a maintenance program, production will probably not be interested and the maintenance department will say you're meddling in their business. However, everyone manages equipment. The operators do, and the maintenance staff does; engineering, supervision and management do. In this way, you can get everyone to buy into an equipment management and maintenance program.

Components of TPEM

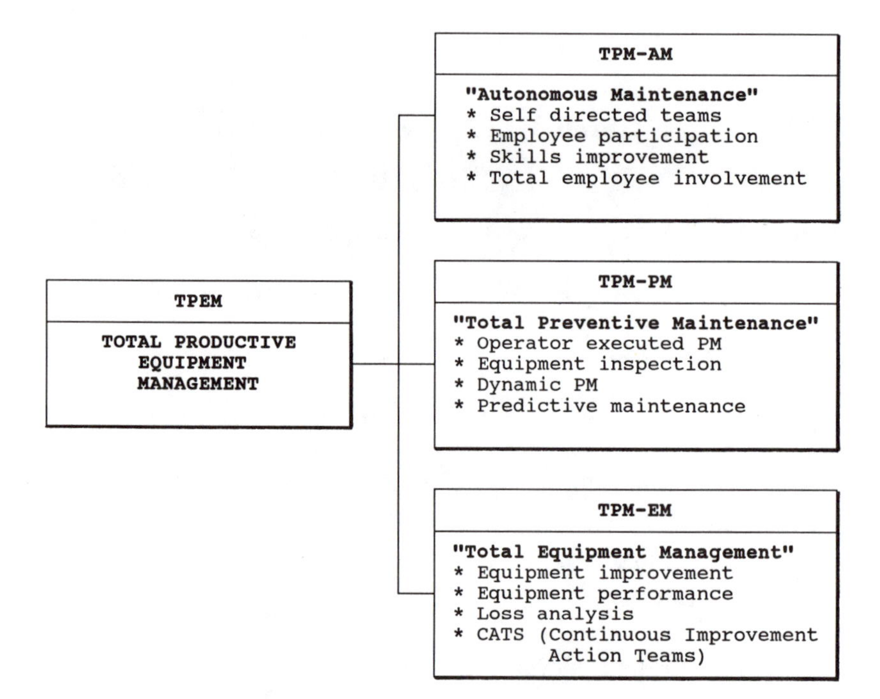

Figure 13

The Components of TPEM

The components of TPEM are TPM-AM, TPM-PM, and TPM-EM (see Figure 13). Once you have analyzed your current OEE's and established your goals, you are in a position to use these tools in the appropriate manner and sequence to help you improve and manage your equipment.

Before dealing with the three components in more detail in separate chapters, a basic overview is given here. This will help in the understanding of the available options and facilitate the approach to customize your TPM plan.

Autonomous Maintenance with a Difference

TPM-AM (autonomous maintenance) is usually different from the Japanese version, especially since within TPEM, each plant is encouraged to develop its own workable approach to autonomous maintenance. The focus is on *participation* by the operators in the execution of equipment maintenance (especially PM), inspections, and eventually the fairly autonomous execution of designated maintenance activities. It is *not* autonomous execution of maintenance by the operating department (not even in Japan).

During the installation of autonomous maintenance (which need ***not*** come first under TPEM), the operators will be trained to carry out the maintenance and inspection activities they are motivated and capable of doing. Therefore, a hallmark of autonomous maintenance is training and skills transfer. As Figure 14 illustrates, there are several levels of operator training under TPM-AM.

The operator with no training, no involvement with equipment and no need for maintenance skills will be a relic of the past in most "world class" companies around the world. Therefore, *all* operators should receive training to obtain basic equipment knowledge and basic maintenance skills. This level of operators could be called Technical Operator 1 (TO/1) and might be compared to High School

OPERATOR TRAINING UNDER TPM-AM

POSSIBLE TITLE:	OPERATORS INVOLVED:		LEVEL OF KNOWLEDGE AND SKILL:	COMPARE TO:
TO/3	SOME	ADV. TRAIN-ING	ADVANCED EQUIPMENT KNOWLEDGE ADVANCED MAINTENANCE SKILLS	POST GRADUATE
TO/2	MOST	SPECIFIC TRAINING	SPECIFIC EQUIPMENT KNOWLEDGE SPECIAL MAINTENANCE SKILLS	COLLEGE
TO/1	ALL	BASIC TRAINING	BASIC EQUIPMENT KNOWLEDGE BASIC MAINTENANCE SKILLS	HIGH SCHOOL
OP		NO TRAINING	NO INVOLVEMENT WITH EQUIPMENT NO NEED FOR MAINTENANCE SKILLS	

Develop training plan for each level, depending on need
Develop training schedule
Develop certification for each level

Figure 14

equivalency. Obviously each plant must determine what should be included in this training and develop its own specific training plan and schedule.

The next level of training is "specific training," and *most* operators should go through that. It provides specific knowledge about their machine and specific maintenance skills (again for their equipment). As before, the training content, plan and schedule must be developed. Operators having accomplished this level could be called TO/2 and may qualify for a higher level of pay under a "pay for skill" plan. It is recommended that operators be certified at each level of skill.

The highest level of training is advanced training, for which only *some* operators qualify. They become the specialists and obtain advanced equipment knowledge and maintenance skills. They may play leading roles in equipment set-up and adjustment, including programming of sophisticated equipment.

Within teams of operators, there is now a variety of skills. Everybody can do the basics, others are good with tools, others can identify and analyze equipment problems and develop solutions, others can set-up, adjust and program equipment. This is the power of teamwork under TPM and will lead to much greater autonomy and involvement of operators in managing their equipment.

An Ounce of Prevention Is Not Enough

TPM-PM focuses on preventive and predictive maintenance. Here is where many companies fall down on the job. A recent survey revealed that about 95% of U.S. companies are not performing PM properly. They are behind schedule, postpone (or cancel) too many PM jobs, don't have a complete system of PM checklists, work orders and schedules, and usually don't have dedicated personnel assigned to perform PM. This includes some of the best known companies in the world, with a generous dose of high tech companies that run expensive equipment.

Why do most companies around the world have such a difficult time doing PM? The emphasis is on putting out fires, fixing breakdowns. PM scheduled for today can be done tomorrow; nothing bad will happen. Except tomorrow, the same situation will repeat, and PM will fall further behind. Before they know it, they are in the next cycle, and their PM did not get done at all.

Because most companies have so much trouble executing PM, there is a vital need to consider an *alternate approach* that will give you a higher percentage of completion for PM tasks. There are two options. Maintenance can improve their PM system and execution (probably with increased staffing). Or operators can participate in the PM, cleaning and inspection of their equipment. Typically, as the operators become more involved with their machines, they will start asking for increased participation in those activities. Of course, the ultimate goal is to eliminate breakdowns, freeing up much time for PM and equipment improvement activities.

Predictive maintenance (PDM) is a different story. It is unlikely that the operators will be able to perform much predictive maintenance. It requires use of sophisticated equipment and a level of training that most companies are not able or willing to give to machine operators. But by transferring more preventive maintenance to the operators, you will give your maintenance staff more time to do predictive maintenance.

Improving Your Equipment

The third component is TPM-EM, Equipment Management/Equipment Improvement. The goal is to improve performance and availability of the equipment and the quality of the product by improving the equipment itself. Depending on the condition and age of your equipment, this could be a major and costly task. But historically, the high return on investment and the major increases in productivity make it a very important (and early) activity of TPM.

The participation of the operators, who run the equipment day after day, along with the maintenance staff, supervisors, engineers and sometimes vendors, in group work is the key element to the success of this activity. Normally, operators are quite willing and motivated to participate in EM, since it will improve "their" machine. It is often quite surprising how *much* the operators contribute to this process.

TPM-EM requires considerable time for problem-analysis and problem-solving training and group work to develop equipment improvements. But it can produce spectacular results, if it is properly organized and managed. In addition, the early and high success rate of this activity will lead and motivate the operators to participate in the other TPM activities.

TPM Installation Strategy

Now, how do you go about installing TPM? Do you do AM, PM and EM at the same time? What's the proper sequence to get off to a good start for a successful installation? The answer is that it depends on a number of factors. But here's a good ground rule to follow. In the U.S., a frequently used approach, particularly in unionized plants, is to start with EM, go to PM and end with AM.

The reason for this approach is common sense. As stated earlier, nobody is opposed to equipment management. But if you try to start with autonomous maintenance, you risk opposition, not only from the operators who don't relate to maintenance, let alone to "autonomous maintenance", but also from the maintenance staff, which may not be willing to "turn over" a portion of their jobs to the operators.

By involving both maintenance and the operators in equipment improvement group meetings, you take the burden off both sides. Everyone loves to offer suggestions, to play a role in managing. You can take advantage of this desire and get everyone working together for a common goal. Once the employees are functioning as a team on one project, it is much easier to transfer this newly found cooperation to other activities. Thus, PM and eventually AM become part of the corporate culture.

For a new plant, however, the emphasis is different. There is not yet a great need for equipment improvement in the early stages of a new plant's operation. You want to start with AM, building good work habits of the new operators. Here you have a distinct advantage, because you can prevent equipment problems by having the operators pay attention to their machines from the beginning. Properly installed in a new plant, AM, supported by PM, will maintain your equipment in near perfect condition.

There are several other factors that will influence the installation of TPM. You must set *priorities* and a *sequence* for your installation, depending on the needs of the plant, equipment and personnel. The feasibility study must come first, because it will give you current equipment performance measurements that will indicate your equipment improvement needs. Other information, such as skill levels and training needs will help you in your installation planning. You also have to make a strategic decision on whether to go for capacity expansion or cost reduction, or even both. Quality improvement needs may also determine how to structure and prioritize a TPM installation.

Corporate culture will often have a bearing on a TPM installation strategy. Are the operators ready to accept and carry out TPM-AM? Sometimes union agreements will not allow production workers to use tools, or only simple ones. Obviously, such a situation will impact your TPM strategy, but it certainly will not make TPM impossible.

In order to succeed with TPM, you need to be pragmatic. Do what works in your plant, with your employees, in your environment. Use the TPEM process and apply its components in the proper sequence to make TPM produce the desired results in your plant.

Top management must understand and support TPM. With no top management involvement, the chances for a successful installation are much reduced. Top management should be involved early and develop a TPM vision, set goals, develop the TPM strategy and policy.

The feasibility study is crucial to the success of TPM. It establishes baselines and gives you the answers you need for installation planning. Good long term results come from a well-conceived installation plan, based on a solid feasibility study and a good TPM strategy.

CHAPTER VII

How Much Autonomous Maintenance Do You Need?

Autonomous maintenance is a key element of TPM. It can be the reason for great success, as in Japan, or it can be a potential stumbling block, as in many non-Japanese companies. Sometimes, managers believe that TPM is the tool to install autonomous maintenance. It is the other way around --autonomous maintenance is *one* of the elements of TPM, albeit an important one. However, the differences between the Japanese work culture and the non-Japanese (especially U.S.) culture make it usually necessary to develop a strategy that is different from the Japanese textbook approach.

Japanese autonomous maintenance has been interpreted as groups of well trained operators executing *all* routine maintenance on their equipment, performing all inspections, as well as executing minor repairs. This interpretation, plus the recommended application of the five S's: seiri, seiton, seiso, seiketsu and shitsuke (roughly translated as organization, tidiness, purity, cleanliness and discipline) has caused a fair amount of resistance in Western countries, less so in other Oriental nations.

So, do you abandon autonomous maintenance? Certainly not. There is too much at stake. The benefits of autonomous maintenance are too numerous for you to abandon it. They are: much better operating equipment, substantial maintenance cost reduction, much less equipment downtime, a highly trained and motivated work force, better product quality and more output.

As a result of autonomous maintenance and their worker's dedication to TPM, award- winning Japanese factories do not look anything like U.S. ones. You will rarely see Japanese maintenance personnel on the shop floor during production shifts. It is somewhat frightening, but the maintenance people don't need to be there. No emergencies arise, no "line maintenance" needs to be there. Nothing goes wrong, nothing breaks down. The machine operators are totally in control. Now the question is: what type, and how much autonomous maintenance will produce results for us? What approach should I use?

Custom-Made Autonomous Maintenance

How much autonomous maintenance do you need? The question should really be the other way around. How much autonomous maintenance can I get? The answer is, as much as you can motivate your workers to do. More is better, of course, but in some plants it is just not possible. For most installations, it is going to be the result of a long, carefully orchestrated training and restructuring period.

Autonomous maintenance is worth the effort. Your equipment will run better because PM, inspections and routine maintenance will be done on schedule. Maintenance costs will go down because much of the travel and waiting time the craftsmen record on each job will disappear. When equipment does go down, it won't be out of production as long, because in many cases the operators will be trained to know what to do to get it back on line.

Not only will your workers be trained, but they will be highly motivated to do the maintenance work. They will understand how their machines work, and will want to keep them running in top condition. Because you're getting more production time from your equipment, autonomous maintenance leads to higher output and better quality, the ultimate goal of all manufacturing.

Recognizing Limitations

However, all these great advantages are not going to fall into your lap. You should be aware of the pitfalls of autonomous maintenance even as you plan to take advantage of its benefits.

You will have to establish the learning ability of your workers. Some will be more trainable than others. More important, you will have to determine if they can be motivated to learn, and what steps you'll have to take to engage their enthusiasm.

How is your maintenance staff going to react to autonomous maintenance? Will they see it as a threat to their job security, dig in their heels, and refuse to cooperate? How will you persuade them to support TPM? And what is the maintenance department's role going to be, once the operators take over routine maintenance and PM? The maintenance department is not going to go out of business under TPM, nor should it. You need to plan to redirect their activities toward new, higher goals, to the "high tech" organization that your plant needs to cope with the new and more complex equipment of the future.

Training *time* is another potential stumbling block. Manufacturing's goal is maximum output from each shift. Production managers are often reluctant to have the workers leave the production area to attend training. Time must be found to execute the training needed for autonomous maintenance. In one plant, the TPM manager, as part of the feasibility study, surveyed all 250 production workers. He found that only two of those workers were so vital and dedicated to the process that they couldn't be taken off the production line. The other 248 were available for training.

You may be competing with other training programs such as quality (SPC), Just-In-Time production, and others. How do you manage that TPM gets its share of this limited resource?

These are some of the handicaps of introducing autonomous maintenance. You have to decide how much "autonomous maintenance" will be right for you. Design and implement a program (TPM-AM) that works best for you.

What tasks can your operators do? Can they clean their equipment? In most cases, the answer will be yes. What about lubricating their machines? You have to remember that they're not automatically going to be able to do it. You may have to color code the grease gun or oil can and the lubrication points. That means training. Can they inspect their machines? If you train them properly, certainly they can, but it will take time.

What about equipment set-ups, adjustments, preventive maintenance, and minor repairs? It will depend a great deal on their trainability, how much they can learn, and their motivation, how much they want to do. The answers in each case will vary according to the work force. There are realistic limitations you will encounter and you have to be aware of them.

Training--The Key to TPM-AM

Success in installing TPM-AM in a plant hinges on training. There is no substitute for this all-important aspect of autonomous maintenance. Because it is so vital, you need to spend considerable time and effort in determining how to go about it. You'll definitely need an understanding between maintenance and production to make it work. And it will require the cooperation of the engineering, training and personnel departments as well.

There are a whole series of questions you must answer before you begin TPM-AM training. First, who is going to do the training? This question could have several answers. The best trainers usually come from the maintenance staff.

These craftsmen know the equipment and the operators. The TPM staff, depending on its size and what other duties are assigned to it, is another good resource. You could have a member of the engineering department do the training, particularly if some member of the staff is an expert in a certain area, such as lubrication. Line supervisors or members of the training department are other possibilities.

Even vendors are often a good resource. They have a vested interest, because if you buy their product, they want to offer training in the proper use of their equipment. Going farther afield, you can investigate the capabilities of local colleges or trade schools to do some training. Just remember that this instruction is probably going to be off-site, more expensive and time-consuming. There are skills training companies that offer such high tech training tools as interactive video disks, programmed instruction and video tapes.

One company had a unique solution to their trainer needs. They brought back retirees on a part-time basis to teach maintenance skills to the operators. These retirees proved to be a very valuable resource, since they had all worked many years in the maintenance department or on the production line, and knew the machines intimately. From this range of choices, you must determine which one is the most practical for your TPM training.

On-The-Job Training (OJT)

Most of your training should be carried out at the work site. It's fast and it's the least expensive way. The training for TPM-AM is often done in short--"one point" lessons that only take 20 to 30 minutes each. Training rooms are used less frequently, but there may be times you want to get away from the noise and urgency of the production line and use a whiteboard or flip charts. One company's commitment to training was so strong they had a complete machine in their training room. A small amount of training, such as attending a seminar or a visit to a vendor, needs to be done away from the plant site.

The time allocated for training will vary according to company policy and the competing demands of other programs. A good rule to follow is to try for one hour per week per operator. Allowing for vacation time and holidays, you will get about 40 hours of TPM training per operator per year with this system. This hour per week is not done all at once, but is broken down into shorter sessions two or three times a week. This approach seems to produce the best results.

You should try to conduct all your training during working hours. Training on overtime is very expensive and can normally be avoided with good cooperation and scheduling. The challenge, then, is to plan when you can take a some time during regular hours to schedule training.

Who is going to plan the training time? It is a function of the TPM staff. They should draw up a plan based on the equipment and operations needs, as well as the time and resources available. This plan should be carefully coordinated with production to ensure minimum disruption.

Levels of Training

As discussed in the previous chapter, not all operators need to be (or can be) trained to the highest level. A mix of skill levels in a TPM small group is perfectly alright. Develop a specific training plan for basic, specific and advanced levels of training (see Figure 14). Adapt the plan to the needs of your plant, the equipment and the operator's functions at the equipment. Existing skill levels (see Figure 15) should be taken into consideration when developing your training plan. Eventually, you should be able to correlate your skill levels with the previously discussed TO/1, 2, and 3 levels of operators. TO/1 may comprise of skill levels 1 and 2, TO/2 include skill level 3 and 4 and the top grade, TO/3 will consist of skill level 5 personnel.

The Cost Of TPM-AM

Unless you do a lot of equipment improvement, training costs are probably going to be your major expense in installing TPM. That's why you have to carefully plan how it is going to be carried out. If you schedule one hour per operator per week, you know your personnel costs. To this figure, you have to add training material expenses.

Before you take this step, find out what training materials may be available from other plants in your corporation. One multinational company with 10 plants held a meeting to discuss TPM. When they began talking about training, one manager said he was about to purchase a skills-training course, with interactive video disks, for $16,000. A manager from another plant said, "Don't do it, I've got that same

TPM	CHART OF SKILL LEVELS - OPERATORS Date: _________
SKILL LEVEL	Description/Attributes/Comments
1	Trainee, basically unskilled; is learning how to operate equipment; unsure of him/herself, needs almost continuous supervision; may be unable to learn.
2	Can operate equipment, knows the basic process. Needs occasional assistance. Does not know equipment very well, rarely recognizes equipment malfunction or quality problems.
3	Operates equipment with confidence and needs very little assistance. Recognizes equipment malfunctions or quality problems, but can not correct them.
4	Knows equipment very well and operates it with a high level of confidence. Needs no supervision. Understands relationship between equipment performance and quality/productivity. Recognizes equipment malfunctions and makes corrections/ adjustments. Could supervise others.
5	Experienced operator who knows equipment and process very well. Supervises and trains others. Highly aware of equipment malfunctions, even of potential problems. Makes corrections/adjustments, inspects equipment and makes minor repairs. Highly aware of equipment condition/quality and productivity relationships. Potential supervisor/team leader.
Approved by:	Industrial Relations ______________ Production ______________ Maintenance/TPM ______________

Figure 15

program sitting in my office. We've finished using it. It's yours." That comment opened the floodgates. Other managers shared what materials they had and by the end of the meeting, the company had a good percentage of the training materials it needed at no extra expense.

That situation contrasts with an international electronics firm that had developed similar training materials independently in about six different plants. That's six times what it should have cost them. It is usually much less expensive to translate training materials than to develop from scratch.

Certification

Another part of training is operator certification. As the workers advance from one skill level to the next, you should document this progress. For one thing, it gives you a very good idea of the skills inventory of your workers.

Figure 15 shows a chart of skill levels, which can be used to assess the capabilities attained by each worker. This chart should be supplemented with more specific training goals during your TPM-AM training phase.

Each operator should be tested or "checked out" after completion of all the courses required for a certain skill level or grade, including the practical application of the learned tasks on the equipment. A certificate should be issued to the operator. A chart for average operator skill levels can be developed for each group, department or the whole plant to document the progress made since the start of the training (Figure 16).

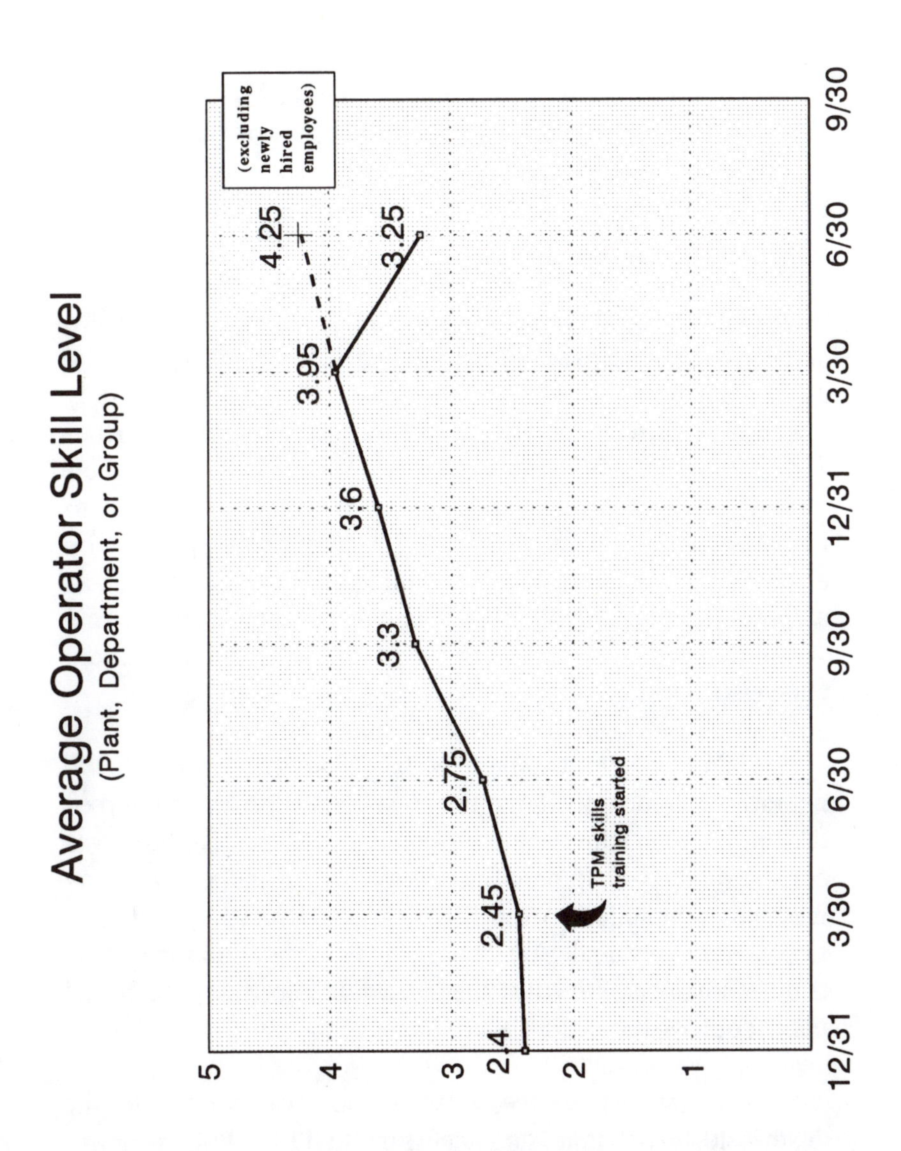
Average Operator Skill Level
(Plant, Department, or Group)
5
4
3
2
1
12/31
3/30
6/30
9/30
12/31
3/30
6/30
9/30
2.4
2.45
2.75
3.3
3.6
3.95
3.25
4.25
TPM skills training started
(excluding newly hired employees)

Figure 16

"My Machine" Concept

Providing incentives to promote your TPM-AM program is an important aspect in the success of autonomous maintenance. Remember, you want to motivate your operators, to instill a sense of pride in them that they know how to take care of their machines. The "my machine" concept is vital to TPM-AM. They may have been involved with improving their equipment under TPM-EM and are starting to participate in the PM of "their" machine. The equipment they "own" looks and functions better and they develop a degree of pride. Now the challenge is to further improve it and "take care of it." Some teams put their names, sometimes even their pictures on their machine to indicate their "ownership." This action should be encouraged, or even initiated by management, as it serves as a great motivational tool. Once the operators assume "ownership" of their equipment, you'll notice a marked change of attitude.

Another plant installed a large board with all the operators' names on the vertical axis and the various types of training on the horizontal axis. As an operator was certified for each type of training, he or she got a shiny metal disc attached to the board. This display served as a skills inventory and as a measure of each operator's and team's advancement, a visible measure that everyone in the plant could see.

There are, of course, many other ways to build the "esprit de corps" that is so necessary to TPM-AM. You will probably come up with your own creative ways to motivate the workers in your plant. The important thing to remember is that this motivation, this pride in skill and ownership of machines, is a vital part of TPM-AM. Without it, your installation will not be as successful.

TPM-AM is an excellent tool for lowering costs and improving productivity. Empowerment of teams works. It does take careful planning, a firm commitment to proper training, and well-motivated workers who want to keep their machines running at top efficiency.

CHAPTER VIII

How to Design and Install an Effective PM Program

Most companies around the world don't do a good job of preventive maintenance. When it is carried out by the same department that responds to emergencies (breakdown maintenance), PM always seems to fall short. However, PM is *absolutely vital* to maintain equipment in top condition.

Therefore, a way must be found that enables your plant to execute all required PM when due and scheduled. Obviously, under TPM, you are looking to your operators to participate in that effort.

Before you start training and involving your operators, you must first establish what type of PM your plant does (or needs) and establish some definitions.

Various types of PM might be:

- Routine (and highly repetitive) PM
- Major PM
- Equipment overhauls
- Predictive Maintenance (PDM)

Another approach, very useful for TPM, is to split applicable PM work into two categories:

Type I: PM work that operators are capable of performing now, or later, after training

Type II: PM work that requires the skill of a maintenance craftsman.

This division of PM work helps in the planning and execution of TPM-PM. Obviously, routine and highly repetitive PM is usually Type I.

It is also necessary to establish a *definition* of PM, as there is very often confusion as to what exactly PM is. Opinions range from the short PMs done "on the fly" to anything (including major rebuilds) that helps prevent equipment breakdowns. Everybody is correct, but it helps to define the various *types* of PM, as the amount of time and the level of skill required for each type varies greatly.

Types of PM

Routine PM could be defined as:

The *systematic*

- Cleaning
- Lubrication
- Inspection
- Testing
- Adjusting, tightening
- Servicing
- Minor repairing

to maintain equipment in perfect running condition.

It is clear that all those activities are likely candidates for Type I (operator performed) PM. Each task normally takes only a few minutes and the travel time for maintenance personnel usually exceeds the actual working time on the equipment. The emphasis here is on *systematic*, meaning that there is usually a set number of tasks for daily, weekly or monthly PM performed the same way at predetermined, repetitive times.

Major PM usually involves:

- Partial dismantling of equipment
- Use of various tools
- Replacement of numerous parts or components
- Higher level of skill
- Much more time than routine maintenance

- Maintenance planner involvement
- Scheduling the equipment for planned downtime
- Test run of the equipment

If we accept this definition, it is obvious that these activities are more likely Type II PM work. However, the equipment is normally not removed from the floor and operator's participation in Major PM is beneficial--what a great way to learn more about "my machine." Food for thought for your TPM-PM planning!

Equipment Overhaul (rebuild) usually involves:

- Removal from the production floor
- Total dismantling of equipment
- Rework or replacement of many parts, components or systems
- Many tools, including machine tools
- Upgrading of equipment
- High level of skill
- Repainting of equipment
- Vendor involvement
- Re-calibration
- Test run
- Re-installation on production floor
- Major amount of time
- Maintenance Planner/Scheduler involvement

It is quite obvious that this is Type II PM work (if you accept this as PM at all).

Predictive Maintenance (PDM) is frequently carried out separately from PM, especially if performed by engineering. However, it serves the same purpose as PM: to *prevent* equipment breakdowns by *predicting* when certain components, such as bearings, gear boxes or motors are going to fail.

Predictive Maintenance includes:

- Vibration analysis
- Megger testing
- Spectrographic oil analysis
- Thermographic analysis
- Infrared testing
- Non-destructive testing
- Expensive test and recording equipment
- Use of computers for analysis and trending

This class of maintenance obviously expands the historic definition of PM, as well as the Type I and Type II definition. But when planning for TPM-PM, you should seriously consider what activities (if only reading of instruments) *could* be Type II or even Type I PM work. There are TPM companies where *operators* read and interpret vibration signatures on computer screens built into the equipment. There are many other companies where the maintenance-based PM staff performs all predictive maintenance tasks.

PM Strategy

Once a distinction between the various types of PM has been made and accepted, you can start planning for TPM-PM. The goal is 100% compliance to the PM schedule, at least on your critical equipment (PM tasks completed versus planned).

There are two approaches:

1. Improve the system, the organization, the execution and control of PMs done by the maintenance department

2. Transfer as many routine PM jobs as feasible to the operators

Under TPM-PM, you should do both. The development and installation of a PM systems for maintenance is discussed first.

An Effective PM System

Step 1: Establish Equipment Data

Step 2: Assign Type of PM and Criticality

Step 3: Develop PM Checklists

Step 4: Develop PM Work Orders

Step 5: Develop PM Routes

Step 6: Develop PM Schedules

Step 7: Maintain Equipment History

Step 8: Apply Bar Code Technology

Step 9: Develop Reporting System

Step 10: Establish PM Organization

Step 1: Establish Equipment Data

Most companies have a computerized list of equipment inventory or equipment data cards available. If not, you should establish data for all your machines that includes:

- Type of equipment and serial number
- Description and manufacturer
- Date of manufacture
- Name plate data (voltage, HP, etc.)
- Updates or changes made
- Location in plant
- Reference to spare parts lists and drawings
- Reference to manuals, etc.

Your machine inventory is your starting point. Most equipment manufacturers recommend inspection and PM tasks for their machines in their equipment manuals.

Step 2: Assign Type of PM and Criticality

Now, you should make some basic decisions regarding PM for each machine. Do you want to include it for operator performed PM (Type I)? Maybe not right away, but later. Or is it a type of machine where operators will not perform any PM? Do you want to include this machine for Predictive Maintenance? Also, at this point, work with production to establish the criticality of the equipment.

For example:

Criticality 1: Equipment failure will shut down the plant or a line, may be a threat to worker safety or may cause harm to the environment. Obviously, you don't want any of this to happen, hence Criticality 1. It means the equipment *will* be inspected and PM carried out when scheduled with no exception and that the equipment *will* be made available by production when scheduled with no arguments. PM compliance must be 100%.

Criticality 2: Equipment failure may shut down a line, may be a potential threat to worker safety or the environment. You can afford to shut down the equipment for a short period of time, because there may be redundant equipment available. PM compliance must be 90-100%. That means that no more than 10% of planned PM tasks may be postponed or canceled.

Criticality 3: Equipment that is not critical to the production process, such as stand alone machines that are not in constant use or have adequate redundancy. PM compliance must be 80-100%, i.e. no more than 20% of planned PM tasks may be postponed or canceled.

The application of criticality enables you to perform the right PM if you don't have time to do *all* your PMs due to a temporary shortage of personnel or a production crisis.

After you have made these decisions regarding your equipment, you can then proceed to develop the various tasks for PM and PDM execution.

Step 3: Develop PM checklists

Each machine has its own specific checklist, typically containing rather standardized tasks, that will appear on many other checklists, such as cleaning tasks, checking for lubrication loss, looking for loose bolts, etc. There may be different checklists for daily, weekly or monthly PMs or one master list may be developed to cover all frequencies.

Normally, PM checklists contain no parts or only simple parts or materials (such as filters or lubricant), which are readily available at or near the machine. Equally, checklists should require only simple (or no) tools. You should estimate the time required to complete each checklist for planning and control purposes. A typical daily or weekly checklist takes only a few minutes to complete.

There should be two types of checklists for each machine. One covers PM performed while the machine is running. There are certain tasks, such as detecting overheating or excessive vibration that can only be done when the machine is running. Other jobs, such as checking the tension of a V-belt or internal cleaning can only be done when the machine is turned off and secured. The goal is to perform as much PM as possible when the equipment is running, in order to limit the time the machine must be taken out of production.

This type of PM (using checklists) is usually Type I work, suitable for operators.

Step 4: Develop PM Work Orders

As opposed to checklists, PM work orders require tools and materials and are normally done by maintenance. Work order PM is also routine and repetitive, but normally at lower frequencies, such as monthly, quarterly, or annually. However, this is not yet major PM.

Each PM work order is also related to a machine and contains a list of tasks and a bill of materials. It may need PM planner/scheduler involvement to plan the parts and materials and schedule the work, especially if running hours, number of hits/strokes/parts made, etc. determine the time to execute this PM. Typically, PM work orders have a specific craft assigned to them and the time required is also estimated.

This type of PM is normally Type II, but operators may be involved assisting in its execution, since the machine is usually shut down.

Step 5: Develop PM Routes

The PM Route is the best tool to improve productivity of the maintenance personnel carrying out PM checklists or work orders. It often takes longer for a round trip from the maintenance shop than to complete the tasks on the machine. The route sheet eliminates the round trips by stringing the PM jobs together in a certain area. The craftsman basically follows a "road map" and moves from machine to machine.

You may be surprised how much PM work can be accomplished using that approach. Since you have previously estimated the time required for each checklist or work order, you now can establish the total time for each route by adding the travel time to the total work time. As with the PM work orders, the frequency (weekly, monthly) is shown on the route sheets and there will be a separate sheet for equipment running and equipment shut down.

Step 6: Develop PM Schedules

Normally, there is an annual schedule for each machine, containing all PM frequencies. This schedule is quite static (nothing changes), unless PM is driven by running hours or other variables. The master schedule triggers the release (daily or weekly) of all checklists or work orders due. Schedules for operator-performed PM are normally posted on the machine or are in a binder nearby. The controlling schedule will be signed off when the PM is completed.

Under TPM-PM, you will find, however, that the checklists and even the schedule become more dynamic, as the feedback from the operators and maintenance staff is reflected in additional, or fewer, tasks and in different time intervals on the schedule.

You can promote good PM compliance by leveling out the hills and valleys of your daily PM workload by

TPM EQUIPMENT HISTORY

Equipment No. *91061812* Description *Mixer* Asset No. *B-27498*

Acquisition date *June 18, 19XX* Cost *$2,500.00* Replacement Cost *(19XX) $3,000.00*

Date	W.O.No.	Description of Action	Labor Hrs	Labor Cost	Parts Cost	Total Cost	Cumul. Cost	% of Repl.
01/22/XX	14721	Replace Gear Box	3.0	60.00	358.00	418.00	418.00	13.9
01/30/XX	14844	PM	.5	10.00	0.00	10.00	428.00	14.3
02/15/XX	14987	Replace Guard over Drive Pulley	1.5	30.00	40.00	70.00	498.00	16.6
02/28/XX	15368	PM	.5	10.00	12.00	22.00	520.00	17.3
03/03/XX	15652	Replace Drive Shaft	2.5	50.00	50.00	100.00	620.00	20.7
03/25/XX	15877	PM	.5	10.00	0.00	10.00	630.00	21.0
03/30/XX	16300	Replace Drive Bearing	1.5	30.00	25.00	55.00	685.00	22.8
04/10/XX	16521	Repair Gear Box	8.0	160.00	30.00	190.00	875.00	29.2
04/20/XX	16854	Align Drive Shaft	3.0	60.00	0.00	60.00	935.00	31.2
04/28/XX	17201	PM	.5	10.00	0.00	10.00	945.00	31.5
05/05/XX	17727	Replace Drive Bearing	2.0	50.00	32.00	82.00	1027.00	34.2
05/17/XX	18221	Paint Equipment	4.0	80.00	10.00	90.00	1117.00	37.2
05/26/XX	18922	PM	.5	10.00	0.00	10.00	1127.00	37.6
06/20/XX	19301	Replace Gear Box	4.0	80.00	360.00	440.00	1567.00	52.2
06/27/XX	19644	PM	.5	10.00	0.00	10.00	1577.00	52.6

Figure 17

developing a good PM schedule. This will permit you to utilize a constant number of maintenance personnel dedicated to PM. It is also important to limit production interruptions by combining, for instance, a monthly and quarterly PM job to be done at the same time, even if one of the cycles needs to be changed a bit.

Step 7: Maintain Equipment History

A good equipment history is vital for equipment management, maintenance and improvement. Unfortunately, only a few companies maintain and use good equipment histories. Without it, you won't be able to pinpoint repetitive failures or establish total repairs costs as compared to the replacement costs. Equipment histories also help you adjust your PM efforts and to develop a good approach for equipment improvement.

Figure 17 shows an example of a good equipment history. Each repair or major PM is added as a one line entry to the history, which is kept for each major piece of equipment. Include the date, work order number, a short description of the action, labor hours and cost, parts costs, total repair cost and cumulative cost. Since all this is normally done on the computer, it can calculate the percentage of cumulative cost as compared to the replacement cost. This will help in the process of making equipment replacement decisions. With no annual repair cost and history available, it will be very difficult to justify equipment replacement and could lead to costly postponement of these decisions.

The following example illustrates why maintaining an equipment history can save you money. A large steel plant had about 5,000 pieces of mobile equipment, ranging from small fork lifts to diesel locomotives. The maintenance manager knew that certain machines were always in the shop and were probably costing a lot of money. The company kept maintenance records, but no equipment history. When a consultant came in, he created equipment histories from those records and showed that the company spent up to three times the amount of replacement cost on annual repairs and maintenance on certain machines. This is the same as your spending $45,000 *every year* to keep your $15,000 car on the road!

Step 8: Apply Bar Code Technology

More and more U.S. companies are turning to bar code technology as a productive and high-tech approach to manage and control maintenance activities. While bar coding is commonplace in supermarkets and many stores and in production areas for inventory control, it is not frequently used in maintenance. It provides numerous advantages, not the least of which is the elimination of the need to write down information, something that many maintenance people don't seem to do very well.

Here is how it works. Each maintenance work order (including PM, even operator executed checklists, if you wish) is bar coded when printed out. The computer knows the contents of the work order. When the maintenance craft starts the job, he or she wands (reads with a light pen) the work order and the personal badge, which is also bar coded. Now the computer knows the job has started, when, and by whom, and moves the job from the backlog into the active file.

At stores, where parts are issued, all parts are labeled with a bar code tag; small parts in drawers have the label on the drawer or in a catalog. The work order is wanded and each part issued against this work order is also read. It does two things: a) it automatically charges the parts and costs against the work order and b) it automatically adjusts the inventory and may even issue a purchase request if the minimum number of stock quantity is reached. At night, when stores may not be staffed, or at altogether unstaffed stores, access to stores is obtained by the bar coded badge and the computer knows, who entered stores and when.

Next, the maintenance worker goes to the machine to carry out the job and wands the bar code tag affixed to the machine. After the job is finished, the work order is wanded again to close out the job. Now the computer has all the information to do the following (without the maintenance craft having to write down *anything*):

a) *Close out* the work order, including the date, total cost and time required and remove it from open work order file

b) Add the job to the *equipment history* (by equipment number) including date, work order number, job description, time and labor cost, cost of materials, total cost, accumulated cost and percent of replacement cost

c) Calculate *PM compliance* (PM work orders scheduled vs. actually done)

d) Issue a list of PMs *not done* during the scheduled period

e) Calculate *performance indices*, if time estimates have been used, such as productivity, utilization and labor performance

f) Calculate *MTBF* (mean time between failure), if the work order is coded for breakdown and operating hours are entered

g) Keep *inventory* up to date, including parts usage, monthly cost of materials, cost of materials by equipment, etc.

h) Report total *labor hours used* by equipment, area, total plant and by type of work (breakdown, planned maintenance, PM, etc.)

i) Produce other, custom designed, *reports*

As you can see, bar coding provides a tremendous range of maintenance management and control tools at a reasonable cost. The technology exists, the software exists, all you have to do is develop the system to fit your needs. The hardest part is to convert maintenance stores to bar coding. But this is a one-time task and well worth the effort, considering the control and visibility you gain.

The attractive part of bar coding is that writing is totally, and paperwork is largely, eliminated. Maintenance workers adapt quickly to this system and generally like it. Management finally gains the tools and data needed to manage maintenance *and* equipment.

Step 9: Develop Reporting System

Unfortunately, many companies fly blind when it comes to good PM management. The absence of usable PM reports is a contributing factor. They spend the majority of their time and effort in responding to breakdowns and PM gets done on an ad hoc basis, with little planning and very few, if any, reports. In such an environment, it is difficult to make progress, let alone produce a turnaround. Don't let that happen to you.

PM takes a lot of commitment and discipline. And the results don't show up immediately. So it takes patience too. When the results do show up, you need to document them to justify your investment in PM and to keep going.

For that reason, there are two types of PM reports. One type tells you how well you are *executing* your PMs and the other tells you how *successful* your PM activities are, with regard to a positive impact on your equipment.

The *control* reports include the following:

- PM compliance (the amount of PM done on schedule as compared to planned)

  Goals: 100% for Criticality 1 equipment
  90% + for Criticality 2 equipment
  80% + for Criticality 3 equipment

- PM performance, utilization and productivity, as discussed in Step 8

- PM costs (labor and materials)
    - by equipment
    - by maintenance and by operators (that ratio will shift under TPM, so you need to track it)
    - contractor cost, if any
    - total cost by departments and plant

- Equipment History, to be used as a tool, not only as a report

The *progress* reports include the following:

- Downtime hours
    - by equipment
    - department
    - plant

The downtime hours are frequently supplied by production

- Downtime trend (as above)

- MTBF (mean time between failure) for each Criticality 1 & 2 machine

- Value of increased uptime, possibly correlated to PM costs

Creating and using these reports help you execute your PM in an organized fashion. If you do your PMs right, and on schedule, downtime will shrink and equipment performance will be up. It is important to measure this progress, as the value of good PM is not obvious to everybody in the plant.

Step 10: Establish the PM Organization

The PM system just described can only be successful if it is supported by a good PM organization. It is *highly recommended* to use a *dedicated* PM staff (i.e. PM specialists who only do PM--and stay on schedule). If you manage to level your PM workload (Step 6), the PM staff can be kept consistent.

And surprisingly, your PM staff is not large under this system (especially when you have your operators participating in TPM-PM). Adding together all estimated times for all PM work orders and maintenance executed checklists (including travel time and allowances) will give you the total labor hours per week. Divide this number by the work hours per week to get the required dedicated staffing. Once you have determined the staffing, you can create the organization structure. Most plants will wind up with a rather simple organization (a small PM group in maintenance). However, large plants may need a PM supervisor for a staff of 10 or more and possibly a dedicated PM planner/scheduler.

The Secrets of Successful PM

It has been established over a period of time that companies that have a very successful PM program have the following elements in common:

- A good *system,* computer supported

- PM *routes*, for PMs done by maintenance

- *Dedicated* staffing
- *Criticality* assigned and followed
- Good *reports* and equipment history
- Absolute *management commitment* to PM

Operator-Based PM

The second approach to your PM improvement strategy is to transfer as many routine PM jobs as feasible to the operators. In some instances it is quite easy to do, especially where operators are already involved in cleaning equipment, setting it up and making adjustments, fixing minor problems, such as jams, etc. They are generally motivated to do more on their machines. Then it becomes a question of what to transfer, and to conduct the necessary training. Later chapters will discuss how to go about that.

In other instances, quite the opposite is the case. The operators are reluctant to "touch" their machine, often because they have been prevented to do so for years by existing company policy or practice. Sometimes labor agreements between union and management prevent operators to use anything but "simple tools."

Operator-based PM presents a unique opportunity to substantially increase your equipment's performance while maintaining (or even decreasing) the total manufacturing budget. If offers new opportunities for participation, involvement and training for operators.

Computer Friendly

It goes without saying that the previously described 10-step PM system must be computer supported.

The PM work orders and checklists are printed out in the planner's office or on the floor. It is practically impossible to produce a manual schedule with so many PM activities going on. The automation of data gathering and the reporting functions is a must. The same goes for maintenance inventory control.

Fortunately, there are numerous CMMS (Computerized Maintenance Management Systems) available that will do a good job of supporting your PM system. Many are PC-based, most support a LAN (Local Area Network) system. For large plants, there are mini- and mainframe-based CMMS programs.

For plants with an existing CMMS, introduction of TPM will bring an additional challenge. You need to include a new "trade", namely the operators, in your scheduling system. The distribution of PM checklists and schedules will be wider and may include additional terminals on the floor. The reporting system must distinguish between maintenance- and operator-performed tasks. More people, including operators, may have to be trained to input PM data.

However, most good PM programs can be adapted without major work. It is important to develop your needs and specifications under TPM first and then adapt or purchase a system that will support the execution and control of your improved PM program. Remember, don't fly blind!

CHAPTER IX

Improving Equipment Through Problem-Solving Techniques

TPM-EM, Equipment Management/ Equipment Improvement is the third component of the TPEM (Total Productive Equipment Management) process. But in many companies, it is the first one that is being applied. There are several reasons for this approach.

As opposed to TPM-AM and TPM-PM, the operators here provide brain power and equipment experience, not manual labor. They participate in teams to analyze equipment problems and to develop improvement ideas. Another reason is that frequently TPM-EM results in quick and significant improvements, which will give you a head start towards a successful TPM installation. Yet another reason is that EM provides an early and non-threatening involvement of operators with TPM and their equipment, building confidence and motivation to go on to AM and PM. And they get to interact with maintenance and engineering, an important step in the process of team building.

CATS

To start the process, the TPM small groups, sometimes called "Creative Action Teams" or "Continuous Improvement Action Teams" (CATS) are organized (usually on a voluntary basis) from the operators working on a specific machine, several similar machines, or from a part of a process or assembly line. About five to seven operators make a good size team. The teams need to be supported by at least one person from maintenance who is familiar with their equipment and by a process, production, manufacturing or industrial engineer who is also familiar with the equipment and can provide guidance during the analysis and improvement process. Depending on your corporate culture and organization structure, the supervisor or coordinator of that area may also participate in the CATS meetings.

The primary goal of the teams is to identify and analyze equipment problems and then to develop solutions and proposals for improvement.

Feed the CATS

The CATS need information and data to be successful. They are dealing every day with breakdowns, minor stoppages, delays, reduced speed, long set-up times and other problems. They may need improved equipment, safer and easier to operate. What are the biggest problems, the largest improvement opportunities? They need input, data.

Figure 18 illustrates the process of the CATS or TPM small group activities. There are four main sources of input:

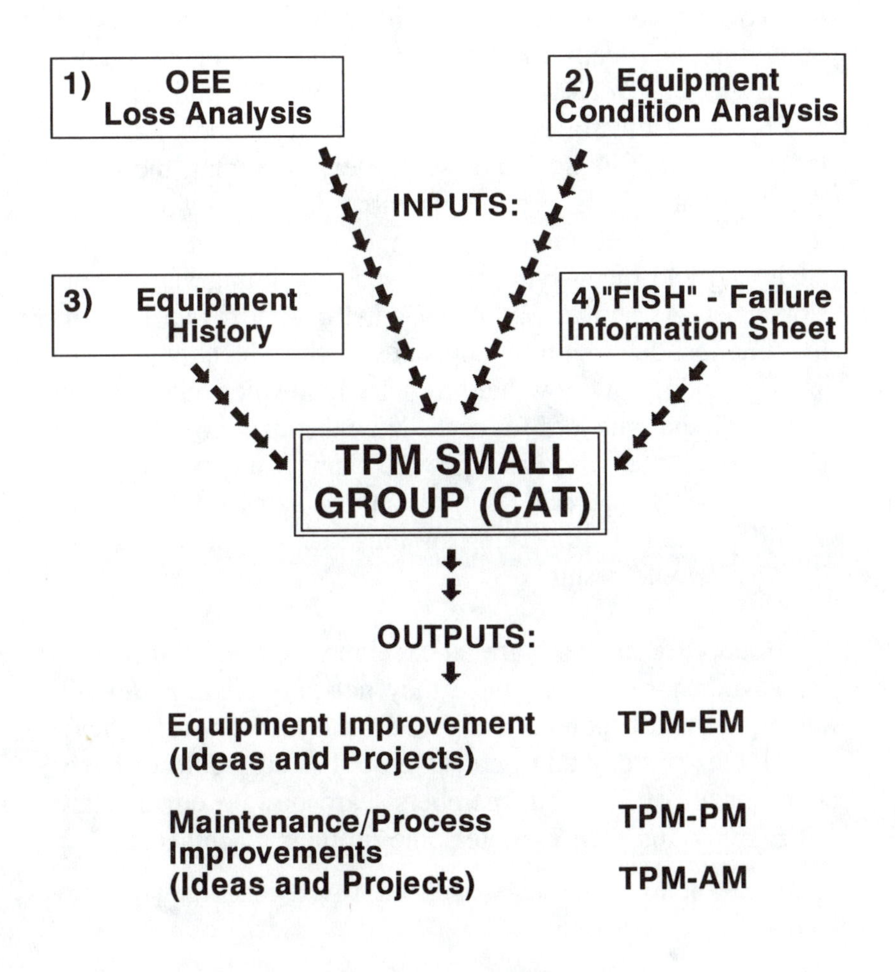
TPM Small Group Activities
1) OEE Loss Analysis
2) Equipment Condition Analysis
INPUTS:
3) Equipment History
4)"FISH" - Failure Information Sheet
TPM SMALL GROUP (CAT)
OUTPUTS:
Equipment Improvement (Ideas and Projects)
TPM-EM
Maintenance/Process Improvements (Ideas and Projects)
TPM-PM
TPM-AM

Figure 18

1) OEE Loss Analysis

This data should be available from the feasibility study, where the equipment was carefully measured and studied and the various losses have been identified and quantified. Sometimes a pareto analysis is available from this study, otherwise the CATS will develop their own. As the example in Figure 19 illustrates, the first level pareto shows a 20% loss due to idling and minor stoppages. Assuming the team decides that this is a good potential for improvement (it usually is!), they may go to the second level pareto, analyzing and quantifying the reasons for idling and minor stoppages. As shown, jams account for almost one half of the time lost due to minor stoppages. A third level pareto, or a short study, may now help to determine where these jams occur and the time loss for each. Now the team goes to work to uncover the *reasons* for the jams and to develop improvements that will eliminate those jams. If the team would succeed in eliminating all jams, a 9% improvement in the OEE would result.

Good OEE analyses are an excellent tools for equipment improvement activities, since they not only clearly *identify* the equipment problems and losses, but also quantify them in minutes per day and percentages. It is advisable to have some or all of the team members learn how to conduct an OEE study and then execute some on their machine(s).

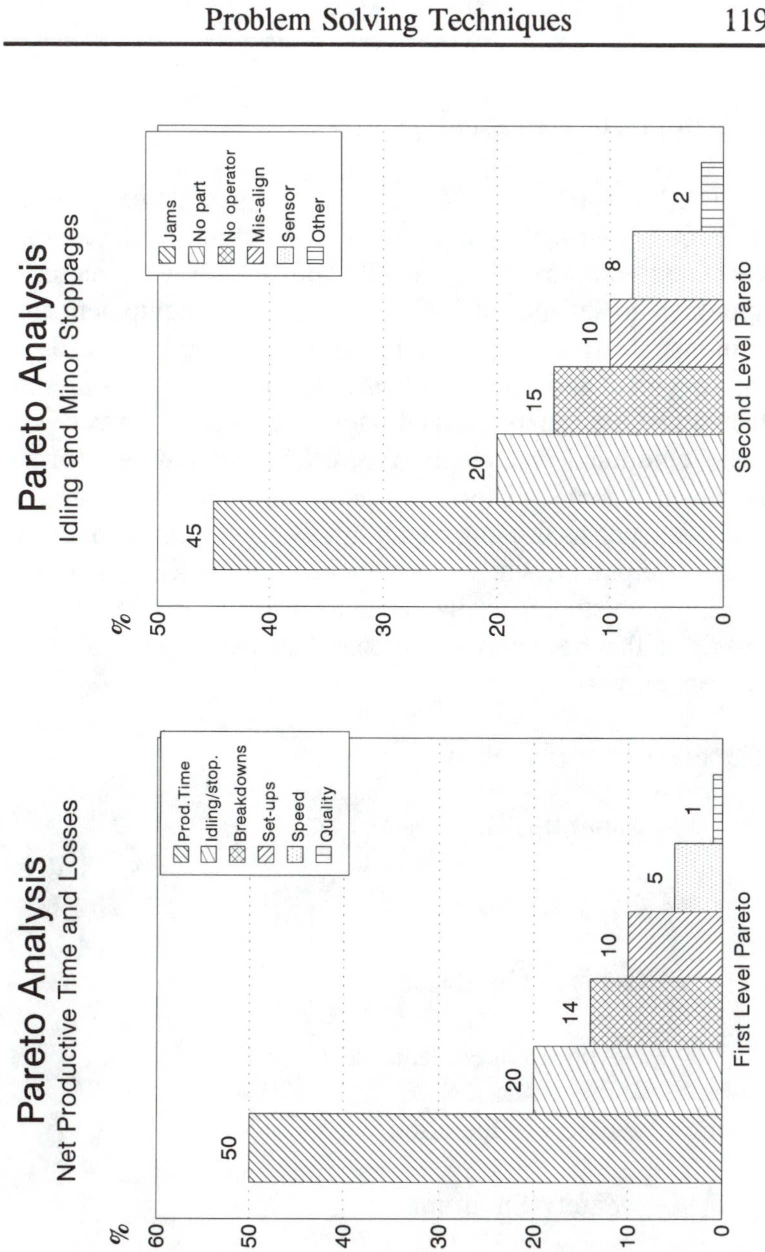
Pareto Analysis
Net Productive Time and Losses
%
60
50
40
30
20
10
0
Prod.Time
Idling/stop.
Breakdowns
Set-ups
Speed
Quality
50
20
14
10
5
1
First Level Pareto
Pareto Analysis
Idling and Minor Stoppages
%
50
40
30
20
10
0
Jams
No part
No operator
Mis-align
Sensor
Other
45
20
15
10
8
2
Second Level Pareto

Figure 19

2) Equipment Condition Analysis

The form in Figure 20 is one of the most powerful tools to involve your operators early (and sometimes with passion) with their equipment and the TPM process. It is a structured approach to let the operators critique their equipment and demonstrate that their equipment may be in need of improvement and better maintenance. It is much less precise than the OEE analysis, and more subjective. Sometimes, quite emotional statements appear on the returned forms. However, some different aspects are included in this analysis, such as missing capabilities or the general condition of the equipment, which is not included in the OEE analysis. Having both studies available as input for the CATS meeting provides the best and most complete assessment of your equipment.

The equipment is evaluated for:

- **Reliability**
- **Capability**
- **General Condition**

    -Appearance/Cleanliness

    -Ease of Operation

    -Safety/Environment

Your equipment may require different or additional criteria.

TPM EQUIPMENT CONDITION ANALYSIS

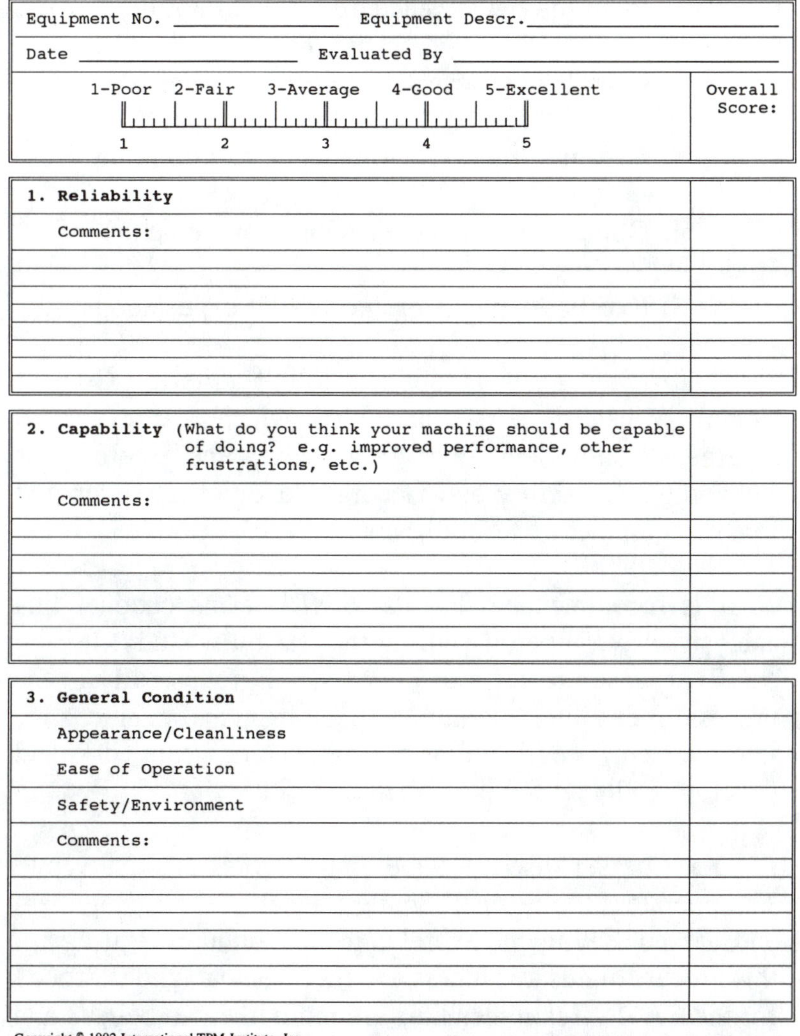

Equipment No. _______________ Equipment Descr. _______________________

Date ____________________ Evaluated By ______________________________

1-Poor 2-Fair 3-Average 4-Good 5-Excellent

1 2 3 4 5

Overall Score:

1. Reliability

Comments:

2. Capability (What do you think your machine should be capable of doing? e.g. improved performance, other frustrations, etc.)

Comments:

3. General Condition

Appearance/Cleanliness

Ease of Operation

Safety/Environment

Comments:

Figure 20

The rating scale is as follows:

1. **Poor** (below all standards, should not be used)
2. **Fair** (barely acceptable, below standards)
3. **Average** (meets requirements, can be improved)
4. **Good** (could be better with improvements)
5. **Excellent** (meets or exceeds all expectations)

This rating scale is applied to all criteria. General Condition has three sub-headings, which are evaluated separately and then averaged. Next, the scores for Reliability, Capability and General Condition are averaged for the equipment's Overall Score.

It is recommended that the CATS teams conduct this analysis early, hopefully during the feasibility study. It is an excellent reason to form the TPM small groups early in the process, focus their attention on equipment and maintenance improvement needs, build motivation and teamwork and develop readiness for participation in TPM.

Don't be surprised if some of the scores are quite low. The operators are normally quite critical of their equipment and are not shy about expressing their opinion. But again, this process builds awareness and involvement, and is "food" for the CATS. Team members are more likely to believe and take action on analyses and data that they themselves have been involved in developing.

Equipment Condition Analysis

Condition - Action

Rating Scale	Condition	Possible Actions
1 POOR	• Below all standards • Very difficult to operate • Unreliable • Very low OEE • Does not hold tolerance • No improvements done • Unsafe to operate • Very high scrap rate • No PM	**Requires immediate attention** • Scrap • Rebuild • Start PM • Improve function and safety • Clean up • Repaint • Hide
2 FAIR	• Barely acceptable • Below standards • Not easy to operate • Limited capability • Dirty • Low OEE • High scrap rate • Very little PM	**Requires early action** • Rebuild • Improve function and safety • Improve PM • Clean up • Improve inspection
3 AVERAGE	• Meets requirements • Fairly reliable • PM done • But not in good condition • Some limited capability • Decent appearance • Average OEE • Average scrap	**Requires action** • Improve necessary functions • Improve inspections • Improve PM • Clean up • Don't let deteriorate
4 GOOD	• Reliable machine • Nice appearance • Very little scrap • All PMs done • Some improvements done • Good OEE • Meets all standards	**Possible actions** • Fine tune PMs • Keep inspecting equipment • Keep cleaning/lubricating • Improve where possible • Don't let deteriorate
5 EXCELLENT	• Perfect condition • Looks brand new • Excellent capabilities • No scrap • Equipment improved • No breakdowns • Perfect PM done • Excellent OEE	**Use as example** • Show off to customers • Don't let deteriorate • Maintain perfect PM record • Keep perfectly clean

Figure 21

The table in Figure 21 links typical conditions found to the rating scale in order to make the scores more consistent from team to team. In addition, possible actions for each condition are shown to prompt the CATS, management, engineering and maintenance to address the need for improving the equipment and maintenance.

3) Equipment History

If the equipment history, as discussed in Chapter VIII, is available, use it for additional input for the CATS. OEE and equipment condition analysis normally don't show repetitive failures or breakdowns, but equipment history does. It is the last piece of the puzzle that completes the total picture of equipment performance, condition and history. It also shows the maintenance and repair cost of the equipment, which should be used in the improvement decision making process.

Very often, a decision to increase maintenance (especially PM) is rewarded by a substantial benefit from increased equipment performance (output) and quality improvement.

4) Failure Information Sheet (FISH)

In many cases, there is no usable equipment history available. Or maybe you want to focus your operator's attention even further on equipment breakdowns. The Failure Information Sheet is the perfect tool for this. It is a form (Fig. 22) that is filled out by the operators every time a failure on their equipment occurs. It asks three rather simple questions and requires some feedback, including speculation.

TPM FAILURE INFORMATION SHEET ("FISH")

Equipment No. ______________	**Equipment Descr.** ____________________		
Date ______________	**Time** __________	**Operator** ____________________	

1. What happened? (Describe breakdown)

2. Why? (What do you think caused the breakdown?)

3. What would you do about it? (To prevent future breakdowns of of the same kind)

Figure 22

As with the equipment condition analysis, be ready for some surprises, but positive ones in this case. Past use of this form has shown that the operators could not only describe very well what happened, but in most cases, they knew exactly what caused the breakdown. And most importantly, many answers to the third question, "What would you do about it?" were extremely usable suggestions that led to quick and significant reductions of breakdowns! If not, it provided discussion material for the CATS meetings that eventually led to improvements. That's why "CATS love FISH"; the Creative Action Teams find the failure information sheets and the information contained in them a very useful tool for their work.

With that much information available, OEE loss analysis, equipment condition, history and failure information, the teams are well equipped to develop numerous and continuous improvements on their machines. Sometimes the question becomes where to start. Obviously, anything that will improve the throughput of a bottleneck machine will have priority.

ROI is another consideration: cost of the improvement versus the benefits. However, most projects that reduce idling and minor stoppages, sometimes even breakdowns, are not that expensive. It is often the lack of proper PM and cleaning that is the cause of the problem, and in that case, the operators can help themselves. For that reason, starting with TPM-EM is often the recommended approach. It motivates the operators to participate in TPM-PM and AM.

CATS Meetings

With suddenly so much to do, it becomes important to carefully plan the meetings in order not to lose too much production time and to accomplish good team progress. The meetings should be scheduled in advance at a regular time, such as every Wednesday morning at 8:00 a.m. and should last about one hour. Of course, frequency and length of meetings will vary according to your needs and the success of the teams. Some teams will be highly productive and develop many improvements, while others will not.

A team leader is crucial to the success of the CATS. He or she must lead and motivate the team members without dominating the discussions. Try different approaches, a lead operator, a supervisor, a maintenance craft or an engineer, to see what works best in your environment. An engineer is advisable in complex technical situations or where a large amount of training or analysis is required, but you may not have enough engineers for all teams. It is quite possible to rotate team leadership, once the team is operating efficiently.

A well-planned agenda is important too. The teams should go through a set program on each item. They should keep minutes of these meetings so there is a record of what actions they decided on, who is going to follow up on these decisions and the deadlines set. You can post copies of the minutes on activity boards around the plants to let other operators and teams know what your CAT is doing and the success that they are accomplishing.

Analyzing The Problems

Your operators may never have had any formal training in failure analysis. You train them in several simple techniques that can help them analyze the possible causes of equipment failure and come up with workable solutions.

The first technique is the *Pareto analysis*, sometimes called the 80/20 rule. It was devised by a 19th century Italian economist, named Pareto, who observed that 80% of the country's wealth was owned by 20% of the people. This rule applies to many situations, including equipment breakdowns. There may be ten major reasons why machines fail, but only two reasons will probably account for 80% of the breakdowns. If you focus on finding those two reasons, you can set priorities for equipment and maintenance improvements.

The Pareto rule is easy to apply. You simply list the major breakdown causes, the number of breakdowns per cause, and the total downtime for each cause. Then develop a bar chart to graphically illustrate your major reasons. While Pareto is good for problem analysis and quantification, it is not very useful in solving these problems.

To search for solutions to your equipment problems, you'll want to teach your operators to use the *cause and effect diagram*, also known as the fishbone diagram (Figure 23). The various problems, connected with material, machine, manpower and method, are placed on the diagonal lines leading to the backbone, which is the effect--equipment breakdown or malfunction. Preparing this diagram

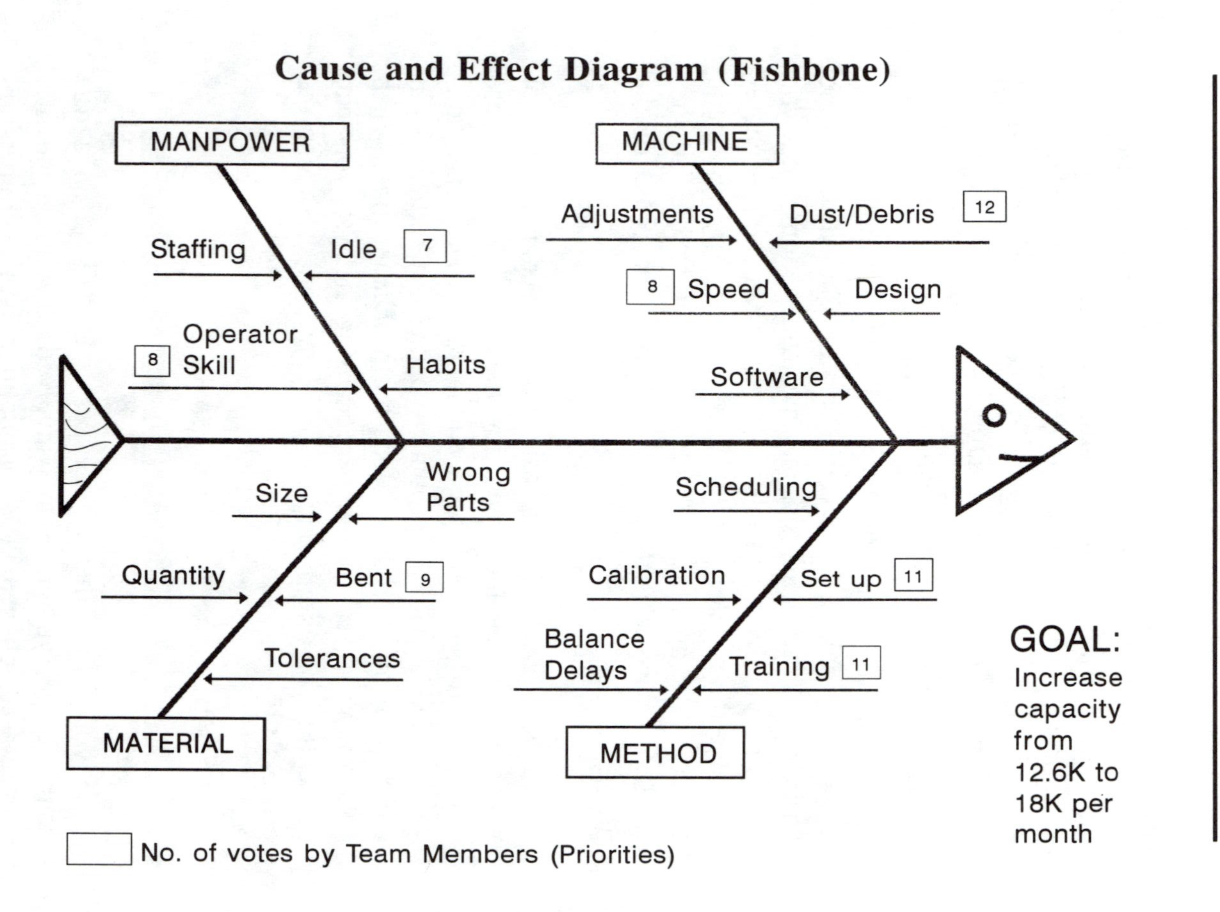

Cause and Effect Diagram (Fishbone)
MANPOWER
MACHINE
Staffing
Idle
7
Adjustments
Dust/Debris
12
8
Speed
Design
8
Operator Skill
Habits
Software
Size
Wrong Parts
Scheduling
Quantity
Bent
9
Calibration
Set up
11
Tolerances
Balance Delays
Training
11
GOAL:
Increase capacity from 12.6K to 18K per month
MATERIAL
METHOD
No. of votes by Team Members (Priorities)

Figure 23

encourages group work to determine causes and its effect on equipment. It also points towards solutions by showing the relationship between cause and effect. After the fishbone is completed, the team often votes on the various causes to set priorities for problem elimination.

A third analytical tool is the *root cause analysis*. It requires a little deeper thinking and is a good tool to use on a specific problem. Here your operators should have some basic knowledge of physics and mechanics. Sometimes you may discover that the problem has more than one root cause. To solve the puzzle, you will have to use a very systematic and scientific approach. You try one solution after another until you narrow it down and eliminate the cause of the problem.

Methods analysis is a basic industrial engineering tool which focuses on the production method. Flow charts are developed to show the production process and what non-value added activities contribute to delays and bottlenecks in the movement of products along the production line. When you know what activities are causing delays, you can concentrate on eliminating the problems and shortening the process flow.

Reaction of Operators

If you follow these guidelines, you will be on a true course to get your operators involved. It is often the first step in a customized TPM installation. By giving operators ownership of the machine and its problems, you guarantee the operators will want to see the problems fixed. They also want to make sure that the problems don't re-occur. Therefore, they will more likely be motivated to participate in equipment management activities, which include increased participation in maintenance activities through TPM-PM and eventually TPM-AM. You will find that the operators and other team members are going to be very proud of their accomplishments.

They can now improve their machines and analyze and fix problems as an independent CAT team, with the help of maintenance. Recognize and reward those accomplishments. Positive reinforcement by management is a strong motivational tool. Have the team make a presentation to management and don't hesitate to take the team out to a company-sponsored lunch or dinner. Certificates of award, including a President's Award have been used successfully to recognize and motivate the teams.

Consider sharing the financial benefits with the CATS. Many companies have an existing suggestion program with monetary awards. You can expand such a program to accommodate improvements made and savings accomplished by *teams*. Be innovative and keep the CATS motivated and excited. Equipment and maintenance improvement activities never stop. True Total Productive Equipment Maintenance is an ongoing process that will put your production facilities into a very competitive world class position as you move towards the 21st century.

CHAPTER X

The Feasibility Study

The approach to TPM varies greatly between the typical Japanese factory and the non-Japanese plants. The reason is part culture and part TPM experience. In Japan, TPM normally comes "top down", i.e. the chairman or president of the company or the plant manager announces the decision to install TPM. Since authority is highly respected, nobody questions that decision or balks at it. TPM training is conducted, the TPM organization is formed and away they go, planning for and executing the installation.

Not so in the typical non-Japanese plant. It's usually a mid-level manager or an engineer who "discovers" TPM. Then it must be "sold" or justified to top management. Part of that justification process, or simply to find out if the plant is ready for TPM, is the execution of the feasibility study.

There is only *one* Chairman of the Board of a large U.S. corporation whom this author knows about, Mr. Harold A. Poling of Ford Motor Company, who, in a corporate-wide policy letter, set forth the concept of "Ford Total Productive Maintenance" and established top management responsibility for interpreting and implementing this policy letter.

But even at Ford, it took the pioneering efforts of managers such as Charlie Szuluk, now General Manager, and engineers such as Michael O'Connell of the Electronics Division (ELD) to create awareness for TPM and launch it in their division first.

There *are* companies however, that just are not ready (yet) for TPM. Sometimes the corporate climate is totally wrong for TPM. On rare occasions, the union turned down TPM, without even waiting for a study. However, that is diminishing now, as the benefits of TPM become better known to everybody. About half the plants in the U.S. where TPM is currently being installed are unionized.

Another reason for the different approach is TPM experience. Practically every Japanese executive knows about TPM (it's been around there for over 20 years) and knows it works. Again, that is not the case in most other countries. You must work for TPM acceptance and fight for staffing and funding. Since these resources are normally not available in abundance and must be applied wisely, a feasibility study to determine the viability and benefits of a TPM installation is quite appropriate. It makes sense under those circumstances to determine priorities (where the greatest need for improved equipment performance exists) and to start in areas that offer the best chances for success. Once TPM is successful in those areas, the rest will take care of itself. It helps to accomplish an early break-even, when benefits start exceeding costs, since this will demonstrate the financial viability of TPM and create the goodwill (and cash-flow) to continue plant-wide.

The Japanese TPM manager does not worry about these financial situations and is not concerned about producing an early break-even. He knows he has total top management backing and they all have the patience to wait for the results, even if it takes three or more years. He does not need a feasibility study. That's why you won't find information about the feasibility study in any previous TPM literature.

Your situation is quite different though. Unless you have money to burn, time to waste and a disregard for your corporate culture and plant climate, you *need* to carry out a TPM feasibility study. The results of this study will be your input to custom design your TPM installation for best results.

TPM will have a profound impact on your operations and corporate culture. Therefore, your installation planning must be based on *solid information* and the actual *needs* determined in the plant. We know that a well-planned maintenance job takes about half as long as an unplanned one. The same holds true for TPM. You can not risk planning such a major undertaking based on assumptions!

Contents of a Feasibility Study

The two sides of the TPM coin are *equipment* and *personnel*. So your feasibility study must concentrate on these elements. Also, your current level of *maintenance* (especially PM) must be assessed, since this information will impact your TPM program development.

The major tasks of a typical feasibility study are:

1) Assessment of your current equipment's performance and condition.

Include all your *important* machines and a representative sample of the rest in this assessment.

a) Equipment effectiveness and losses.

Here, you measure the true effectiveness of your equipment and determine and quantify your equipment losses, as discussed in Chapter V. About the only way you can do that with any degree of accuracy and credibility is to *observe* the equipment over a period of time, using the "OEE Observation and Calculation form", as shown in Figure 24. Trained observers will perform *time-studies on equipment* and note the amount of time lost (in minutes) in the appropriate column of the form. The minimum observation period is four hours and, depending on the equipment's cycle time and frequency of losses, longer periods (up to 24 hours) of observation may be required. Even then, it may not be possible to get a good picture of the breakdowns and you may have to research breakdown or maintenance records. But normally, quite a solid picture starts to emerge between four and eight hours and a longer study will not substantially change the percentage of losses observed. Using the numbers obtained from the observation, you can now calculate your OEE, NEE and the percentage of each individual loss, broken down into various reasons in the areas of idling and minor stoppages, and breakdowns.

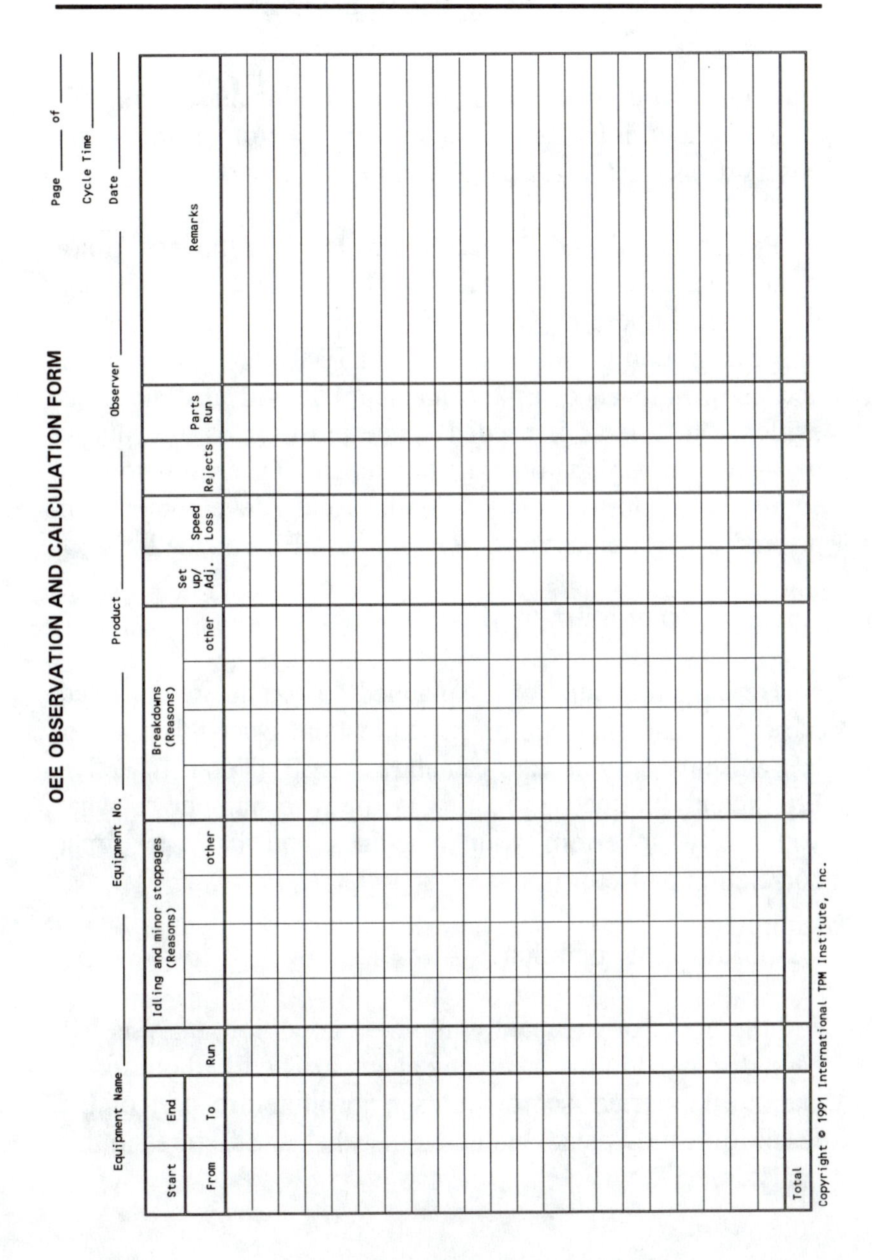

OEE OBSERVATION AND CALCULATION FORM

Equipment Name ______ Equipment No. ______ Product ______ Observer ______ Page ____ of ____ Cycle Time ______ Date ______

Start	End		Idling and minor stoppages (Reasons)				Breakdowns (Reasons)				Set up/ Adj.	Speed Loss	Rejects	Parts Run	Remarks
From	To	Run				other				other					
Total															

Figure 24

Obviously, you must determine those various reasons and fill in the header information on the form before starting the observation. Just ask the operators or do a short test.

Performing the OEE observations is by far the most time-consuming task of your feasibility study. But it is also by far the most important tool you have when it comes to improving equipment under TPM-EM and determining needed maintenance for the machine. It will also serve as evidence that TPM is needed, since most OEEs turn out to be *much lower* than anybody thinks. Determination of equipment improvement potential and development of priorities will be based on your initial OEE calculations.

b) Equipment utilization.

Here the numbers for the planned downtime are collected in order to calculate equipment utilization. Once this number is determined, you can calculate TEEP (Total Effective Equipment Productivity). This is the *true* number of what you really get from your machines and has a direct correlation to Return on Assets (ROA).

c) Equipment condition.

The operator-executed equipment condition analysis has been discussed in the preceding chapter. The filled out forms and the numerical score for each machine are part of the feasibility study data. Include tools, dies and fixtures in this analysis.

2) Assessment of your personnel.

Include all operators, maintenance personnel and first line supervisors. Usually, the personnel department is quite involved during this analysis, as this is a "delicate" issue in many plants. If it is too difficult to obtain all the data, back off. Take a good reading of available skills and in particular the motivation of the employees and close out the feasibility study.

a) Establish skills required.

Develop a list of typical tasks that an operator is currently performing on his/her machine and add potential future tasks to be done under TPM (Figure 25). Using the chart of skills discussed earlier (see Figure 15) determine the *required* level of skill for current and future task requirements for the operator. This list is *equipment* related, not operator related.

b) Establish skills available.

On the same form, add the operator's names (or code numbers) and enter the current level of skill, using the same chart of skills. The current level of skill is provided by the supervisor or is determined by interviewing/observing the operator. In the next column, enter the difference, if any.

c) Skills analysis

Now you can analyze the completed form. It will give you a very good and quantifiable indication of training requirements. A large delta in the total column by task will

TPM Skills Required/Available Analysis

Equipment name ______________________ Date ____________

No. ________________ By: ____________________

Tasks	Skill req.	Op.No. ______	Δ	Op.No ______	Δ	Op.No. ______	Δ	Total Δ
a) Operational								
------	4	3	1	2	2	4	0	3
------	3	3	0	2	1	4	0	1
------	4	4	0	3	1	3	1	2
b) PM/Cleaning								
------	4	2	2	2	2	1	3	7
------	3	2	1	3	0	2	1	2
------								
c) Other activities								
------	5	3	2	3	2	4	1	5
------	3	3	0	2	1	3	0	1

Totals			6		9		6	21

Figure 25

indicate *task*-related training needs. A large delta in the total column by operator will indicate which *operators* need additional training. The grand total for each machine will assist you in determining training requirements by *equipment*. The same approach and process will be used for maintenance personnel (technicians), and, if appropriate, for working supervisors or team leaders.

d) Establish Level of Education

This input is usually provided by the personnel department and is not that critical for the feasibility study. A high level of skill and motivation can be exhibited by people with little formal education. However, many companies do keep statistics on education of its personnel and, if available, makes a good addition to the TPM baseline.

e) Determine Trainability

Since training is one of the major activities of TPM, an indication of (primarily operator) trainability is useful. However, it is not that easily developed. The best input is the successful completion of previous company-sponsored training courses and other courses (such as trade school). Feedback from the supervisors, part of whose responsibilities is training, should also be sought.

f) Determine Level of Motivation

This is an important indicator of the potential success of your TPM installation, but hard to measure. Sometimes a survey, containing several "what if" questions mixed with "job satisfaction" questions has been used. Quite a good approach is to conduct *structured group interviews*, which can also be used to explain the concept of TPM to the group before the discussion. Determination of the *attitude* (towards the job or company, to take a challenge, etc.) can also be accomplished in these group sessions.

g) Employee turnover

In companies where turnover is high, determine the turnover rate for each department and document it for your baseline. TPM normally reduces the rate of turnover and you should be able to document it. Sometimes the turnover rate has a direct association with the level of motivation. The personnel department will provide the numbers.

3) Assessment of Maintenance Effort and Results

This is done on all Criticality 1 and 2 equipment.

a) Current Maintenance Assessment

It is important to determine how much, and what type of maintenance is currently done on your equipment.

Collect information on the following functions:

- Cleaning
- Lubrication
- PM
- Inspections
- Predictive Maintenance (PDM)

- Other planned maintenance
- Breakdown maintenance

and get answers to the following questions:

- Are checklists or work orders available for this task?
- Is a schedule for this task available?
- What is the percent compliance (or completion)?
- Who executes this function?
- Is a report available?
- How much time is spent on breakdown maintenance?
- How much time is spent on PM and PDM?
- How much time is spent on other planned maintenance?

Similar to the skills analysis, the goal is to determine and quantify any deficiency, which then should be addressed as part of the TPM installation. An organized approach to this assessment is shown in Figure 26. It is also equipment related, since improved equipment maintenance and equipment performance are the main goals of TPM. Develop the form to fit your equipment and your current (or planned) tasks. The maintenance department must participate in this evaluation.

b) Proposed Maintenance

Based on the just-concluded maintenance assessment, it is quite easy to develop the proposed maintenance for each machine. This may include higher percentages of compliance of PM activities already done; recommendations for additional checklists, schedules and reports needed; and

Current Maintenance Assessment

Equipment name ______________________ **Date**______________________

Equipment number ______________________ **By:**______________________

	Tasks	**List Available**	**Schedule Available**	**% Compliance**	**Done by**	**Report Available**	**Remark**
1.	Daily Cleaning	√	n/a	70%	Ops.	no	Need better routine
2.	Weekly Cleaning	√	no	60%	Ops./Maint	no	Unclear distribution
3.	Lubrication	no	no	75%	Maint.	no	Need schedule
4.	Daily PM	√	n/a	60%	Maint.	no	Ops. want to do
5.	Weekly or longer PM	√	√	60%	Maint.	yes	
6.	Inspection	no	no	?	Maint.	yes	Need procedure/schedule
7.	Predictive Maintenance	√	√	50%	Engr.	yes	Maint. wants to do
	Etc.						
a)	Estimated % of time spent on breakdown work:					80%	Reduce!
b)	Estimated % of time spent on PM/PDM work:					5%	Too little
c)	Estimated % of time spent on other planned maintenance:					15%	Need more
	TOTAL					**100%**	

Figure 26

additions of tasks not currently performed. This proposal does not have to be final, but will give you a good indication of activities needed under TPM. First opportunities for increased operator participation usually become evident here as well.

c) Commitment to PM

A low percentage of PM compliance often indicates a lack of commitment to PM. There is quite a difference between lip service and commitment. Is there a dedicated staff devoted only to PM? Or are emergencies totally overwhelming the maintenance department? Are there PM controls that monitor the execution of PM? Is the system complete? Findings of low PM commitment and compliance indicate an urgent need for TPM-PM.

d) Equipment Histories

Determine if equipment histories are available, at least for your important machines. If yes, in what form are they and are they used for equipment analysis of repetitive breakdowns?

e) Maintenance Management

The following topics should be briefly assessed and evaluated:

- Maintenance request procedure
- Work Order system
- Planning and Scheduling
- Use of standards or estimates

- Inventory system and stores
- Maintenance Controls and Reports
- Use of CMMS

With or without TPM, a good maintenance management system is essential for efficient and productive execution and control of maintenance.

f) Maintenance Organization

Document the current maintenance organization and staffing, primarily for inclusion into the baseline documentation.

4) Determine State of Housekeeping

There is a direct correlation between housekeeping, productivity and quality. A dirty plant with dirty equipment and no discipline does not produce a high quality product at a high level of productivity. How do you measure up? Are your floors clean, aisles marked and kept free? Are there dirt and grease spots on everything, including your maintenance people's (and sometimes operator's) clothes? An assessment of those factors will raise the level of awareness in your plant. Maybe you need to address housekeeping before or as an early step of TPM.

5) Assess Corporate Culture

Corporate culture and plant climate will have an impact on the success of TPM. Are there existing teams? Do they work? Is there a team spirit in your plant, a high degree of cooperation? Do managers know what the word "empowerment" means? Can you sense a level of enthusiasm, of people involvement? Are there standards of excellence, quality goals, such as Q1 at Ford, Six Sigma at Motorola, etc.? What does "world class" mean to your employees?

Answers to these types of questions will give you a good "feel" of your company's corporate culture. If the answers are mostly negative, you need a "spark" that TPM can deliver, but it takes more motivational work. If the answers are positive, TPM will fit right in.

6) Develop Costs, Benefits and ROI

This step is less and less frequently required, since top management is usually quite aware of the need and benefits of TPM before the presentation of the feasibility study results. However, it is still recommended to estimate the costs and benefits in order to establish goals for a TPM installation.

The cost items include the following:

- Training time
- Training materials development
- Equipment improvement costs (hard to estimate during the feasibility study)
- TPM staff costs
- Meeting costs (during TPM-EM)
- Public relations (PR) for TPM

The benefits include the following:

- Cost reduction
- Productivity improvement
- Downtime reduction
- Postponement of new equipment purchases
- Fewer rejects and less rework

To calculate ROI divide the annual TPM benefits by the annual TPM costs. For instance, benefits are $500,000, costs (investment) are $200,000; ROI equals 250%.

7) Baseline Documentation

This report documents your "current situation" and serves as a *baseline,* against which you measure your progress and improvements. Surprisingly, this step is sometimes forgotten, making it very difficult to calculate or demonstrate your much improved situation two to three years from now.

All the data for this document has been developed during the feasibility study. The baseline documentation is a *summary*, a graphic or tabular presentation of the following data:

a) Equipment effectiveness and losses (departments, plant)
b) Equipment utilization and productivity
c) Equipment condition
d) Skills required/available analysis
e) Current skill level (total)
f) Level of motivation

g) Turnover
h) Current maintenance
i) Proposed maintenance
j) Maintenance management
k) Maintenance organization
l) Housekeeping
m) Corporate culture

Organization of a Feasibility Study

It is obvious that this study will take some resources and time. The best approach is to form a *Feasibility Study Team*, made up of representatives of the whole plant, or at least all the departments where you plan to introduce TPM. Depending on the size of your departments, one team per department may be appropriate. The feasibility study team(s) will be *dissolved* after the conclusion of the study. Its members will return full time to their regular work assignments and typically become leading TPM proponents within their respective TPM small groups (CATS).

It is normally better to have a larger team with part-time members than a small full-time team. It is easier to "borrow" somebody for eight to sixteen hours a week and you'll have more "missionaries" in the plant when the TPM installation starts.

The minimum recommended team size is:

- Two operators
- Two maintenance personnel
- One engineer

- One team leader (typically an engineer or supervisor)

The team is supported by:

- The TPM manager
- Personnel Department
- Engineering
- TPM Consultant, where appropriate

If your plant has a union, it is strongly recommended to invite the union to participate in the feasibility study. Typically then, somebody from the union leadership will attend the training session, some team meetings and of course, the presentation of the study results to management.

Normally, team members are recruited by management, but they participate on a voluntary basis. Make sure they know enough about TPM before you ask them to participate.

Execution of the Feasibility Study

The first step is to *train* the team members in all the activities they will perform during the feasibility study. Very often, a TPM consultant is used for that purpose. Start with an explanation of the purpose, content, organization, schedule and other details of the study. The training and practice in how to perform the OEE analyses will take about one day.

It is recommended to schedule two full days at the beginning of the study to get the team properly trained, allow for enough practice, make the assignments and cover all tasks to be done, including explanation of all forms used. This phase will get the team "off the ground".

Develop the procedures before you begin the actual study. The union should be informed first through normal channels. The employees should be informed by a team member or the team leader. Explain what the OEE analysis is all about (the equipment!) and invite the operator of the equipment to participate by identifying the losses that occur during the study. The supervisor of that area must be informed before team members show up studying his/her equipment and using a stopwatch to measure equipment losses.

Before the team gets busy with OEE and other analyses, identify the equipment that will be studied. If you have five identical machines, maybe only two or three need to be studied; the best, the worst and the average. It is incredible though, how often *identical* machines of the same age have widely varying OEE results.

Once the equipment is identified, you can make assignments. It is better to assign team members to equipment they are familiar with; it makes it easier to recognize and identify losses and their reasons. Each completed worksheet should be turned in daily to the team leader, who will typically input the data into the computer to perform the OEE calculations and summaries.

A schedule should be prepared to plan and control the execution and progress of the study. An example is shown in Figure 27. The total time required varies, depending on

Feasibility Study Schedule

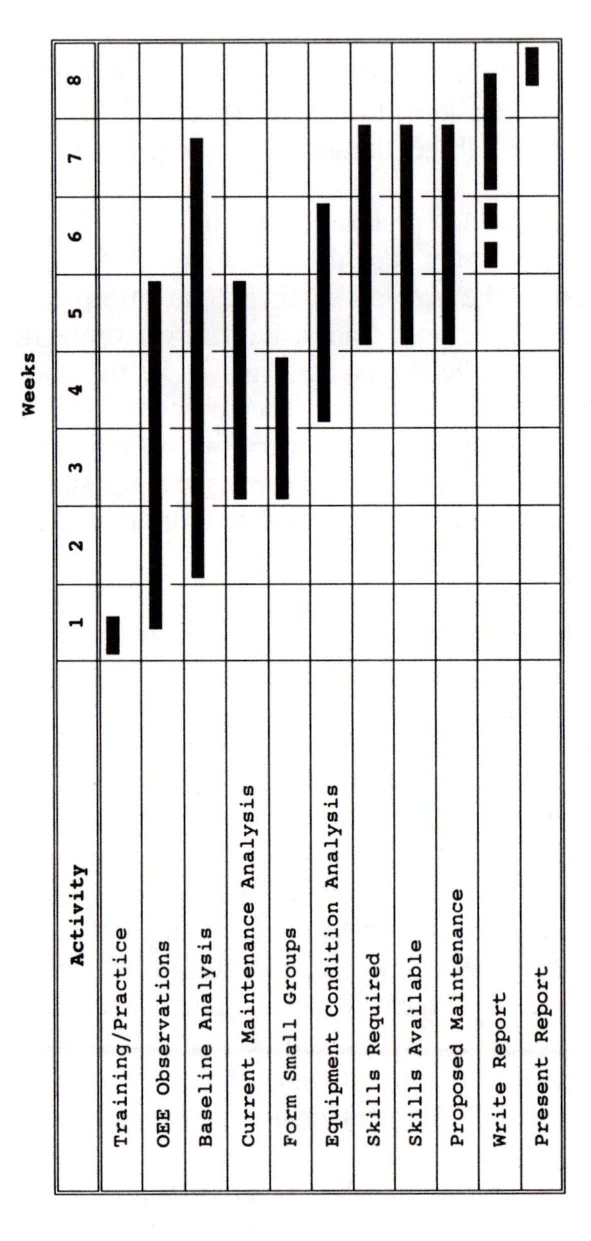

Figure 27

the size of the team and the size (number of machines) of the plant or department. However, an eight-week time frame has been found ideal, allowing time for forms (such as equipment condition analysis) to be filled out, returned and analyzed.

If your company is rather certain that TPM will be installed, it is highly advisable to form several TPM small groups *during* the feasibility study and let them, as a team, complete the equipment condition analysis. This first exposure to TPM, and the corresponding focus on their equipment, usually helps demonstrate the need for TPM; and the process of building motivation has begun.

The feasibility study teams should meet at least once a week to review progress, discuss problems and plan the next steps. About half-way through the study, team members should start thinking about the report and who will develop what portions of it.

Feasibility Study Report and Presentation

The feasibility study will develop so much new and often startling data, that the presentation of the report to management frequently turns into a major event. This meeting often becomes the kick-off of TPM, as there is absolutely *no doubt* that TPM is sorely needed and will produce the hoped for results.

a) *Report preparation*

All team members should participate in the preparation of

the report, but it is usually the team leader, with the help of the TPM manager, who is responsible for the writing and production of the report. The structure of the report should be as follows:

1) Title page, including date
2) Table of contents
3) Management summary (two pages maximum)
4) Purpose of study
5) Introduction, including list of all team members
6) TPM vision and goals
7) Methodology used (approach and schedule)
8) Summary of results (see baseline documentation)
9) Typical examples
10) Other observations, including interview results
11) Conclusions
12) Recommendations
13) Appendix
    a) Equipment analysis details (OEE's, losses, condition)
    b) Calculation formulas
    c) OEE analysis form (examples)
    d) Equipment condition analysis form (examples)
    e) Failure Information Sheet (examples)
    f) Skills analysis details
    g) Skills required/available form (examples)
    h) Current skill level details (by equipment/ department)
    i) Chart of skills
    j) Level of motivation details (from interviews)

k) Current maintenance details
l) Proposed maintenance details
m) Maintenance management audit results
n) Maintenance organization charts
o) Housekeeping assessment details
p) Corporate culture assessment details

The printed report presented to top management usually contains items 1-12 and only pertinent, supporting, examples in the appendix. The full report can become quite voluminous and is used by the TPM staff, maintenance and production managers and appropriate sections by the TPM small groups.

b) *Presentation to management*

This has turned out to be a major event in many companies, attended by up to 40 people. Since TPM has such an impact on a company, *all* managers attend. Of course, the whole feasibility study team is there, plus representatives from the union. Schedule about two hours for this meeting, and announce it far in advance, so everybody can mark their calendars.

Typically, Sections 4 through 12 of the report are presented, using overhead transparencies. Often the TPM manager presents Sections 4 and 6 and the feasibility study team leader presents Sections 5, 7 and 8. All or most team members should take turns in presenting Section 9 (typical examples from the areas they studied). This is frequently quite dramatic, as operators and maintenance people, some of whom have never spoken in public, address the company president or plant manager, whom they may have never met in person. Their presentations are usually very impressive and get the point across loud and clear.

Sometimes, feasibility study teams start to practice TPM-EM, because their analyses show immediate improvement opportunities, which are too good to pass up or postpone. Some teams have presented (or even completed) improvement projects worth hundreds of thousands of dollars. Often, those projects are approved on the spot.

The study team leader then presents sections 10 through 12, always with the strong recommendation to proceed with the TPM installation. Usually during this meeting, management authorizes the installation of TPM, if it has not done so prior to the feasibility study.

CHAPTER XI

The TPM Installation

The TPM installation consists of three distinct phases:

I Installation Planning and Preparation
II Pilot Installation
III Plant-wide installation

Phase I: Installation Planning and Preparation

This is a crucial phase that will greatly influence whether your TPM installation will move along smoothly and with success or whether it will be a struggle. It has been established that an unplanned maintenance job takes about twice as long as a planned one; the same holds true for a TPM installation.

However, if you have executed a complete and successful feasibility study, your installation planning and preparation should be quite easy and will not take too much time.

Depending on the size of your organization, an eight to sixteen week time frame is common.

The typical steps of the installation planning and preparation phase are as follows:

Step 1: Develop installation strategy

Step 2: Develop and put in place the TPM organization

Step 3: Develop TPM vision, strategy and policies

Step 4: Develop TPM goals

Step 5: Conduct TPM information and training

Step 6: Carry out public relations (PR)

Step 7: Develop the master plan

Step 8: Develop plan for pilot installation

Step 9: Develop detailed installation plans

Step 10: Present to management

Step 1: Develop installation strategy

The analysis of your corporate culture, plant climate, levels of skill and education, degree of motivation and, in particular, the *needs* of your equipment and production will determine your installation strategy. Here, you develop your *sequence* of installation. For example, you may do TPM-EM first, followed by TPM-PM, and TPM-AM last.

The above sequence of installation works best in most existing Western plants, as this appears to be "the path of least resistance." In addition, there is usually a need, sometimes an urgent need, to quickly improve equipment. A strong momentum, along with substantial productivity gains, can be accomplished early in the program by starting with TPM-EM, paving the way for the other TPEM components.

A new plant (getting ready for production with all new equipment and new employees) should start with TPM-AM and TPM-PM, and add TPM-EM later. Sometimes the job descriptions for operators to be hired are changed after management decides to introduce TPM, to reflect the different role of a "TPM-operator." There also exists the need to develop a good PM system early (sometimes overlooked), to assure that the equipment stays in perfect condition. With the right strategy, all routine PM, along with cleaning, lubrication and inspections, will be done fairly autonomously by the operators in the new plant.

As it is done in Japan, it seems to be easier to start with TPM-AM in Asian countries, since operators appear to be more willing to follow management direction and respond to the more "disciplinarian" concept of the five S's and autonomous maintenance.

In addition to the sequence of installation, the *priorities* need to be considered for your strategy. It is unlikely that you will be able to install TPM in all areas at the same time. Consider the needs of your equipment and of production and recognize that early improvement of "constraint equipment" will improve the throughput of the whole plant.

Priorities may also be determined by the simple fact that in some areas (or departments) of your plant, the operators and maintenance personnel are quite ready and motivated to "go for TPM", while other areas may not be as ready yet.

Step 2: Develop and put in place the TPM organization

Figure 28 shows the flow of TPM development and the organization needed to execute and support the installation. As discussed earlier, the key function is that of the *TPM Manager,* in some companies called the TPM Coordinator. This position should be established and staffed as soon as management decides to proceed with TPM, or at least before the feasibility study starts. The typical TPM Manager is an engineer or manager with a fair amount of experience in maintenance *and* production. This person should be a good leader and motivator, must be a good communicator and get along with all company employees. The selection of the TPM Manager is probably the single most important decision in TPM development, as this function has the most direct impact on the quality and success of your TPM program.

The TPM Manager usually reports to the Plant *TPM Champion*, who should be a high level manufacturing executive, such as the Manufacturing Manager, who has both operations and maintenance reporting to him. The TPM Champion has executive responsibility for the TPM

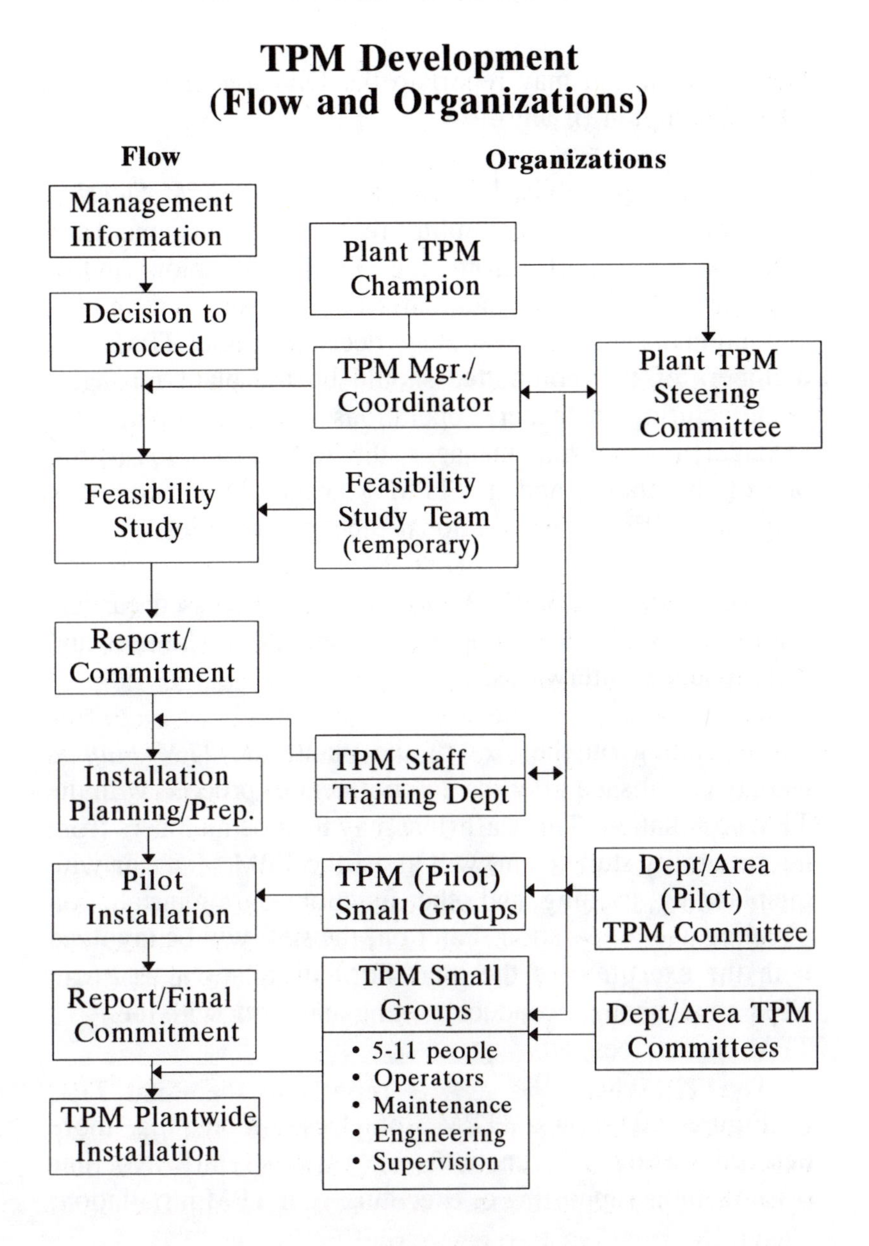
TPM Development
(Flow and Organizations)
Flow
Organizations
Management Information
Decision to proceed
Feasibility Study
Report/ Commitment
Installation Planning/Prep.
Pilot Installation
Report/Final Commitment
TPM Plantwide Installation
Plant TPM Champion
TPM Mgr./ Coordinator
Plant TPM Steering Committee
Feasibility Study Team (temporary)
TPM Staff
Training Dept
TPM (Pilot) Small Groups
Dept/Area (Pilot) TPM Committee
TPM Small Groups
• 5-11 people
• Operators
• Maintenance
• Engineering
• Supervision
Dept/Area TPM Committees

Figure 28

Development and may report to the Division or Corporate TPM Champion (if any).

It is strongly advised to create the *Plant TPM Steering Committee* early. This committee represents management and receives the reports from the TPM Champion and/or Manager. It acts as an advisory group and makes appointments and policy/strategy decisions about TPM. The members of this committee should be the plant manager, manufacturing manager, operations manager, personnel manager, maintenance manager, the TPM Champion (if not one of the above) and the TPM Manager. Your company may have different or additional functions or titles.

The *Feasibility Study Team* has been discussed earlier. During the study, it is supported by the TPM Manager and it is dissolved afterwards.

Depending on the size of the plant, the *TPM Staff* is usually established after the commitment to proceed with the TPM installation. This staff (that may include members from the feasibility study team) will assist the TPM Manager with the planning, training, and other functions during installation planning and preparation. Later on, the staff will be involved with the execution of the total TPM installation, assist all TPM small groups, conduct training and work with the Area TPM committees, etc.

Figure 29 shows "TPM Job Descriptions," a more detailed listing of functions of various staff and line organizations supporting or executing your TPM installation.

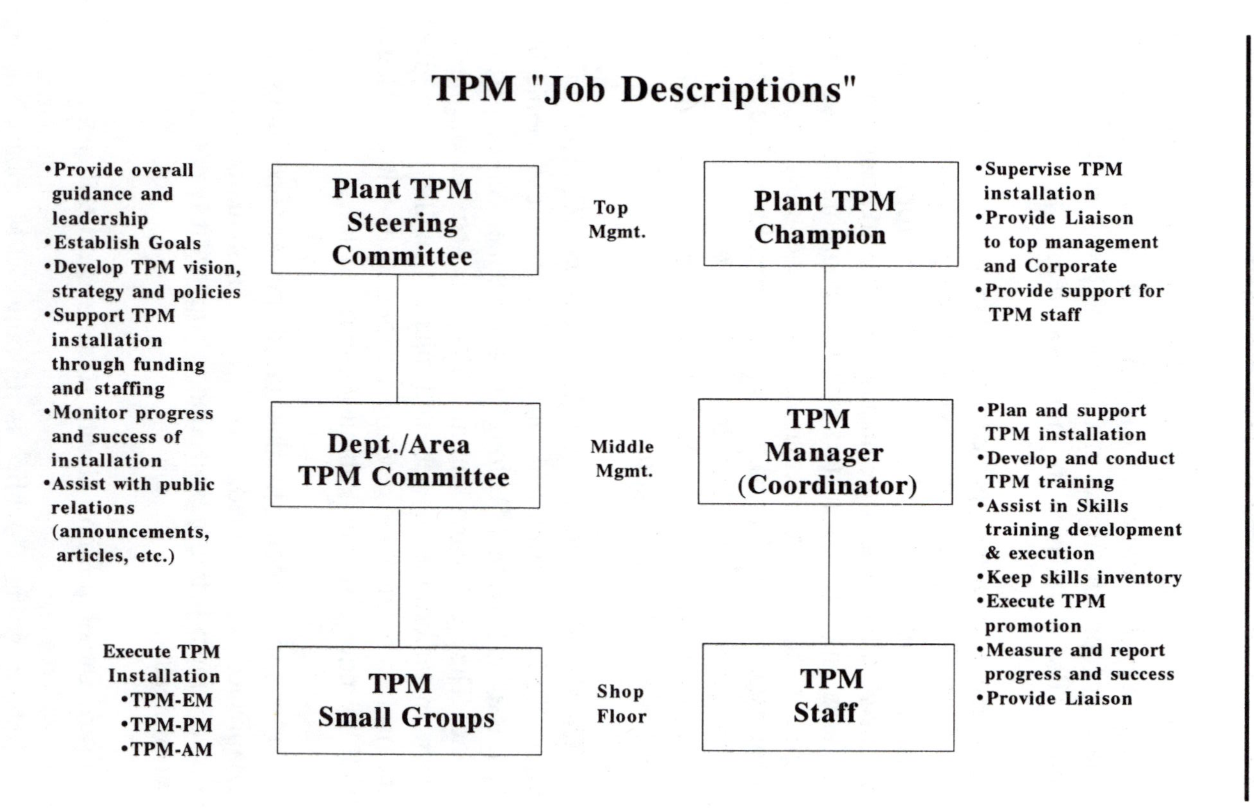
TPM "Job Descriptions"
•Provide overall guidance and leadership
•Establish Goals
•Develop TPM vision, strategy and policies
•Support TPM installation through funding and staffing
•Monitor progress and success of installation
•Assist with public relations (announcements, articles, etc.)
Plant TPM Steering Committee
Top Mgmt.
Plant TPM Champion
•Supervise TPM installation
•Provide Liaison to top management and Corporate
•Provide support for TPM staff
Dept./Area TPM Committee
Middle Mgmt.
TPM Manager (Coordinator)
•Plan and support TPM installation
•Develop and conduct TPM training
•Assist in Skills training development & execution
•Keep skills inventory
•Execute TPM promotion
•Measure and report progress and success
•Provide Liaison
Execute TPM Installation
•TPM-EM
•TPM-PM
•TPM-AM
TPM Small Groups
Shop Floor
TPM Staff

Figure 29

Before you start with the pilot installation, the *TPM (Pilot) Small Groups* (or CATS) need to be established. These groups (discussed in Chapter IX) will carry out the actual TPM installation in your pilot area.

These teams are supported by their own *Area (Pilot) TPM Committee*, which will act as an advisory group and decision making body for the pilot area. The area (or department) manager, the maintenance manager (or the area supervisor), the TPM manager (or a staff member/ coordinator), an area engineer and a representative from the operators and maintenance typically make up the Committee. The same organization holds true later for all *Department or Area TPM Committees*, sometimes called steering committees, during plant-wide TPM installation.

There will be many *TPM Small Groups* (CATS) during your TPM installation and into the future. These should be "natural" groups that "belong" to a machine, cell or part of a line. They will carry out all TPM functions according to your installation sequence and plan. Usually, they give themselves a name and develop a strong team culture. There will be more discussion about those teams later.

It is quite important that all teams and positions be established as outlined. Your TPM installation will be slowed down if the organization to support it is not in place when needed.

Step 3: Develop TPM vision, strategy and policies

Before going "public" with any TPM plans, it is necessary to develop your TPM vision and strategy and establish your TPM policies. The vision should be broad and ambitious, reflecting where your company wants to be five or ten years from now. It is often tied to "world class excellence," quality, partnership, or similar issues and may include a slogan that can be used during the TPM promotion. The TPM vision is developed by the Plant TPM Steering Committee, based on suggestions from the TPM Manager or TPM Champion.

In large corporations, where many or all plants are scheduled to adopt TPM, it is advisable to issue a corporate-wide TPM vision and policy to set guidelines for all divisions and plants. An excellent and laudable example is Ford Motor Company, where Policy Letter No. 18, signed by Ford's Chairman of the Board, Mr. H. A. Poling, establishes the vision and goal as "achieving and maintaining world class manufacturing excellence" through TPM (see Figure 30).

Such a policy statement serves a as clear signal to all plants that top corporate management understands the value of TPM, has made a commitment, and will strongly encourage and support implementation of TPM in all plants, subsidiaries and affiliated companies.

However, it does not absolve each plant from developing their own TPM goals; quite the opposite. The best results have been accomplished where the feasibility study has been done properly, and division, plant and area management have made a strong commitment to TPM.

The TPM strategy must be developed by steering committees. What are your company's manufacturing

Policy Letter No. 18 June 15, 1992

Subject: Ford Total Productive Maintenance

This Policy Letter sets forth the concept of "Ford Total Productive Maintenance"--a manufacturing philosophy and process improvement approach to maximize equipment safety, effectiveness, and life cycles while supporting other manufacturing and product quality objectives. This philosophy and approach are designed to complement the Company's Mission, Values, and Guiding Principles and the concept of Total Quality Excellence.

The goal of Ford Total Productive Maintenance is to improve our overall competitiveness by achieving and maintaining world class manufacturing excellence. A full and sustained commitment of both Company management and individual employees to the implementation of the Total Productive Maintenance process is essential to realize that goal. This is particularly important since the experience of other companies suggests that as much as three to five years may be required to fully implement the process.

The fundamental elements and principles of Ford Total Productive Maintenance are to:

. Significantly improve facility and equipment operation and maintenance processes throughout the Company with the integrated involvement of the union leadership, the entire manufacturing work force and the management organization.

. Improve product quality via improvements in equipment capability and reliability.

. Maximize the overall effectiveness, performance and safety of our manufacturing processes, systems and equipment in terms of productivity, product quality, housekeeping, and elimination of all forms of waste and losses (e.g., equipment breakdowns, product defects, scrap and work place injuries).

. Gain optimum equipment reliability and life cycles and minimize equipment repair and replacement costs.

* * * * * * *

The President and Chief Operating Officer, with the assistance of a Total Productive Maintenance Executive Committee, chaired by the Vice President of Vehicle Operations, is responsible for interpreting and implementing this Policy Letter. Ford subsidiaries and affiliated companies are encouraged to adopt a similar policy.

H. A. Poling
Chairman of the Board

Figure 30

strategic goals? Cost reduction? Capacity expansion? Quality excellence? Just-in-Time manufacturing? You can structure your TPM development to support your overall strategy, and this will impact your priorities or sequence of the TPM installation.

The steering committee must also develop policies before the TPM training and promotion can take place. During the feasibility study, a lot of questions will be raised when people first hear about TPM. Will participation on a team be mandatory? Will TPM result in staff reductions? Will I get more pay if I improve my skills and my job changes? What if I don't feel comfortable doing "maintenance work" on my machine? Will we get a monetary award if we (the CATS) produce substantial cost reductions?

Those, and many other questions (mostly social issues) must be addressed and resolved before informing and training your employees. It is recommended that you make participation in TPM voluntary. Employees who are forced onto a TPM team may display a negative attitude that can destroy an existing good team spirit. An 80% rate of participation is considered a success in Japan. To encourage participation, you may announce that there will be no staff reduction, demotions or pay reduction as a result of TPM. Of course that does not apply if your business volume shrinks dramatically and forces lay-offs.

"Pay for skill" plans are becoming more popular and fit perfectly into a TPM environment. Even without such a plan, most companies have several job classifications, which sometimes are expanded to incorporate operator skills improvement under TPM. As will be discussed later, no function should be transferred if the employee feels uncomfortable about it or feels he/she can't do it safely.

Very often, there is an existing suggestion plan that offers monetary awards to employees who develop productivity improvement ideas. There is nothing wrong with amending and enlarging such an existing plan to accommodate team-developed improvements.

Step 4: Develop TPM goals

This should be relatively easy after a good feasibility study, since you now know your current OEEs and losses. Very often, a goal of 85% OEE, or a 50% improvement over current OEE, is established. There are a variety of other goals and targets that should be established:

- Increase MTBF to a certain level (reduction of breakdowns)
- Increase TEEP to a certain level
- Reduce the rate of defects
- Improve PM compliance to predetermined levels for each type of criticality
- Accomplish a certain percentage of participation in TPM (number of teams established)
- Increase the number of suggestions made by individuals and teams
- Reduce the number of accidents
- Reduce set-up times to a certain level
- Increase average skill levels to a certain point
- and more

Your baseline analysis will offer ample opportunities to establish goals. Develop *target dates* for each goal and do it in realistic steps and increments, not just a global goal for three years from now. Part of your data based TPM implementation will be to measure each area's progress towards the goals.

Step 5: Conduct TPM training and information

This is part of "conditioning your organization" for TPM. Introduction of TPM means quite a culture change for most plants. It is important that everybody involved understands what TPM is, how it works, and how it will affect them. There are several levels and types of TPM training and information:

1. **Management training**
2. **Employee information**
3. **Employee training**

Since management support and commitment is *essential* for the success of TPM, *management training* becomes very important. Often this training is done by a TPM consultant, since nobody in the plant knows yet about all aspects of TPM or about approaches that other plants around the world have used successfully. It is vital to accomplish a "critical mass", which is 50% plus of all management personnel understanding and *supporting* TPM. Experience shows that almost all managers support TPM, once they clearly understand what it is and what it can do. Sometimes, only top plant management receives TPM training and information before the feasibility study is started. Then, it is important to train all other management personnel, right down to the supervisors or coordinators, and the union leaders (if you have not done so before) during this phase of installation planning and preparation.

Typically, this training takes one-half to one day and is conducted in seminar form in-house. The result of the feasibility study and the now established goals and TPM policy should be included in this training.

Basic *TPM employee information* must be given to all operators, maintenance personnel, engineers, members of the training and personnel departments and other appropriate personnel. It can be in the form of a short video tape, a prepared lecture or by other means. These sessions are normally held with groups of about 20-30 people and last one hour. The presentations should take about half an hour with another half an hour left for questions and discussion. Usually the TPM manager or a member of the TPM staff will conduct these information sessions, with the appropriate (and supportive) area manager present. If you are unionized, it is recommended to have a member of the union leadership present as well.

The *employee training* must be given to the small groups when they are ready to start TPM. This training is more detailed and includes a description of the TPEM process and its components (EM, PM, and AM). Details from the feasibility study, the planned approach and sequence of installation, the supporting TPM organization, the goals and policy and more should be covered. This training, also provided by the TPM manager or staff, lasts two to three hours (including discussion) and normally serves as the first step to get the teams launched.

TOTAL PRODUCTIVE MAINTENANCE IN QUALITY MANAGEMENT

What Is TPM?

Plant Maintenance is joining in partnership with each TEC operating division to improve product quality, reduce waste, and improve TEC's state of maintenance. This partnership is called Total Productive Maintenance, or "TPM." The word "Total" emphasizes that all employees will be participating in the maintenance of their equipment.

TPM at TEC is made up of five concepts. Concepts are built on operators and mechanics working together to understand how their roles interact and what they must do to support one another.

1. Utilization of operators to perform certain routine maintenance tasks of their equipment. Operators assuming ownership of their equipment will help to eliminate the potential cause of failure. By taking care of dust, rattles, loosened bolts, scratches, deformation, and wear, all of which combine to cause failures, the operators can do their part to prevent failures.

   The operators will be properly trained and certified to perform the specifically identified tasks. Also, they must have the proper tools to do the jobs.

   Safety must be foremost in the decisions to increase the operators' skills.

2. Utilization of operators to assist mechanics in the repair of their equipment when it is down. Frequently, several pieces of equipment fail at the same time, and PMD does not have enough trained forces to respond expeditiously on all the failures. Sometimes the operators have to be sent home because of down equipment waiting on maintenance. Under this concept, operators would be properly trained to assist maintenance personnel in the repair of the equipment. In return, the maintenance force would be enlarged, the operators would not lose pay due to lack of work, and ultimately the failed equipment would be returned to service much quicker.

3. Utilization of mechanics to assist operators in the shutdown and start-up of equipment. There are times when operators need assistance in shutting down and/or starting up the equipment but do not have the help. This prolongs the shutdown, causing maintenance to wait on the job. By utilizing properly trained mechanics to help operators get the equipment shut down, the outage time of the equipment will be reduced.

   Also, once the mechanics finish the repairs, they would assist the operators in returning the equipment to service by correcting leaks and other mechanical or electrical problems as it is being brought up. By staying at the job site and assisting until the operators have the equipment running, many repeat calls will be eliminated, and overall downtime will be reduced.

4. Utilization of lower bracket labor to perform routine jobs not requiring skilled craftsmen. There are many routine tasks at TEC that could be done by just about anyone, with very little

Figure 31a

training. Under the TPM program, these tasks will be defined, and if it is not feasible for operators or mechanics to do them in their spare time, lower bracket people will be used to do them. The people doing the tasks could report to either operations or maintenance.

5. Utilization of computerized instrumentation to do calibrations using operators or lower bracket personnel. Quality Management at TEC requires that instruments be properly calibrated. Using SPC charts to control operations is based on using data that is as accurate as possible. As part of the TPM program, PMD has purchased a computer and calibration test unit. This system will enable TEC to be more effective by routinely checking and monitoring the calibration of critical instruments.

Pilot Approach to TPM

To evaluate TPM at TEC, in February of 1987, various teams will be established between operations and maintenance to:

A. Select pilot areas where TPM can be tried and evaluated.

B. Select which of the five concepts discussed above will be incorporated in each pilot area.

C. Determine tasks to be studied under TPM.

D. Determine required data and measures to monitor tasks.

The pilots will run for approximately six months, at which time they will be evaluated for improvements and expansion to other areas.

December 1986 article in the *Plant Maintenance Division (PMD) Newsletter* of Tennessee Eastman Company (TEC)

Figure 31b

This information and training process consumes the majority of the time of a typical TPM planning and preparation phase. It is an important part of your TPM development and should not be short-changed, as you are developing a new corporate culture.

Step 6: Carry out PR

Part of preparing your organization for TPM is to carry out some public relations activities. The most common form is articles about TPM in the company or plant newsletter. Figure 31 shows an article written by Bill Maggard in 1986, introducing TPM at Tennessee Eastman Company (TEC). Bill Maggard has proceeded to lead the first, largest and by far the most successful TPM installation outside of Japan, maybe including Japan.

Another excellent, more recent introduction of TPM to plant employees has been written by Andy Gill, TPM manager of the Enfield/Treforest plants (U.K.), which are part of Ford's Electronics Division. These two plants (both unionized) have made excellent and rapid progress, attributable to the following factors: Strong corporate and divisional support (and challenge); very strong and enthusiastic support by the area manager, Paul Taylor; excellent participation by the two local unions; involved leadership by the TPM manager; first rate execution of the feasibility studies by two hard working teams; serious commitment to TPM training; good installation planning and finally a good level of PR and communication, as evidenced by this newsletter article (see Figure 32).

Other PR tools are posters and banners, widely used in Japan and in other Asian nations, not so much in the West. However, the use of TPM activity boards as a PR and in particular, communications tool is becoming very popular everywhere, and it works.

FORWARD WITH T.P.M.

DAVE Boerger has made his first visit to Enfield Plant since taking up his appointment as General Manufacturing Manager of Electronics Division.

And while he was on site, as part of the P.M. Excellence presentation, he was talked-through the Enfield/Treforest 24-point plan to challenge for Total Productive Maintenance (TPM).

He has a heavy involvement with TPM - as the Electronics Division's executive manager who heads-up TPM and PM Excellence as Champion.

So, following in the wake of Q1, and the recently gained PM Excellence, exactly what is TPM?

The first introduction of employees to Total Productive Maintenance came at a conference, spread over two days, which was attended by trade union representatives, managers, staff and hourly personnel.

The seminar was conducted by Edward Hartmann from the International TPM Institute, USA.

Operators

Setting the scene Mr Hartmann described TPM as *"a philosophy which can permantently improve the overall effectiveness of equipment with the active involvement of the operators."*

There are a hat-trick of direct goals to aim for:

- Zero unplanned equipment downtime.
- Zero product defects caused by equipment.
- Zero loss in equipment speed.

Other manufacturing plants around the world which have taken TPM on-board have found that people gain more pride in their work and get greater job satisfaction. Teamwork and individual skills have improved, there is more equipment ownership and the work environment is very much better.

Enfield/Treforest Total Quality Excellence co-ordinator Andy Gill told the Review: "TQE and TPM are distinctly related, as they both target similar goals. A purpose of TQE is to provide a method which will improve the quality of everything we do and this embraces machines and equipment.

"TPM is therefore a vital key in our bid to gain TQE - in the same way that P.M. Excellence played a major part in TQE."

TPM is both cost effective and profitable. It covers problems associated with *all* equipment and not just constraints. It also provides the environment and opportunity for more participation by all employees.

Cleaning

Within TPM there are three main phases. These are AUTONOMOUS MAINTENANCE, a self-governing maintenance programme that can be carried out by setters and operators, like lubrication and cleaning.

PREVENTIVE AND PREDICTIVE MAINTENANCE is what both sites are already doing, although predictive maintenance is a method

• Andy Gill introduces the Shop floor presentation.

which needs to be better utilised and uses indicators like Statistical Process Control, to predict when maintenance is needed. The third phase, EQUIPMENT MANAGEMENT or EQUIPMENT IMPROVEMENT involves a study of present machinery to establish how it needs improving. Feasibility studies in the three phases will start early next year.

"Following on from the studies, Ed Hartmann will assist us again, in putting together an installation plan for TPM which will seek to quickly improve the condition and performance of our equipment," said Andy Gill.

• Dave Beorger (left) meets Machine Shop Group Leader Wally Shea.

Figure 32

Carl DiPasquale, TPM Coordinator of Ford's North Penn plant, came up with beautiful blue, oval (of course) TPM lapel buttons, which were given to every employee after completion of the TPM introduction training. This plant also developed an excellent short video for their TPM introduction, prominently featuring many operators and maintenance personnel who participated in the feasibility studies.

As these examples illustrate, PR activities are quite important in preparing an organization for TPM implementation. It creates a degree of anticipation and enthusiasm, which is needed for active participation and good progress of TPM.

Step 7: Develop the TPM master plan

The TPM Master Plan is typically developed by the TPM Manager and staff. It is a macro overview of the major TPM activities along a time line. Many master plans cover a three-year period.

In Step 1, you should have developed your installation strategy and sequence in general. For each department, decide when you want to start every major activity and how long it should take.

It is difficult to estimate the time requirements for your different activities, as you have no experience values yet. For that reason, the master plan serves primarily the purpose of visualizing your overall activities along a "time line" and in particular the staggered start for the various departments in a large plant (see Figure 33). Once you have a better feel

Example of TPM Master Plan

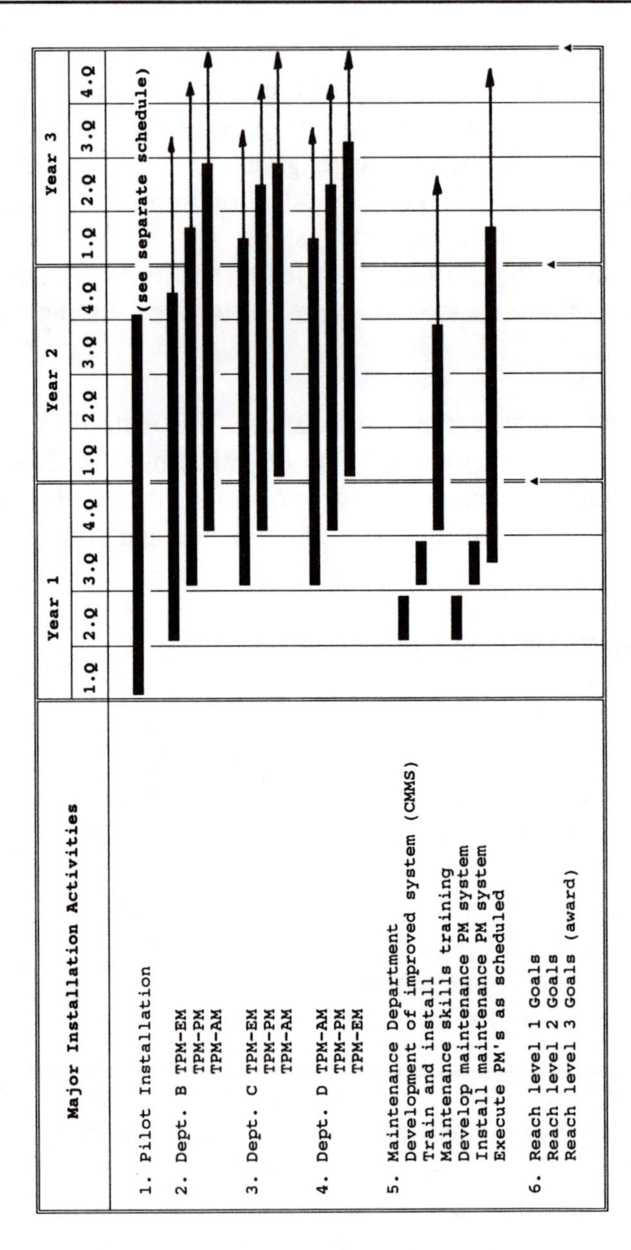

Figure 33

for the actual time requirements during the pilot installation, revise the master plan to make it more realistic.

Don't dwell on the master plan, since at this point you are working with best guesses. Most master plans in existence came straight out of somebody's book and have little relationship to reality. And the good ones, that show a lot of detail, have been developed after the fact. It's a good tool to show the planned sequence of your TPM installation and an *estimated* time requirement.

Step 8: Develop plan for your pilot installation

The pilot installation is a crucial element of your TPM development. Here, you must demonstrate to the rest of the plant that TPM "works." You are testing approaches that have never been done before in your plant. However, the pilot installation allows you to make corrections to your TPM approach before you expand plant-wide. Develop a plan as detailed as possible, based on your installation strategy and priorities (see example Figure 34). The creation of the teams (the CATS) as early as possible is a *key* element in this phase.

The training in problem analysis and problem-solving techniques comes next, if you are starting with TPM-EM. Then, the teams will start to develop equipment and process improvement ideas. Depending on the condition of your equipment, this activity can last for a very long time.

If you are starting with TPM-AM, you may begin with an "initial cleaning of equipment" phase, followed by improvements to reduce or eliminate the source of the problems.

Example of TPM Pilot Installation Plan

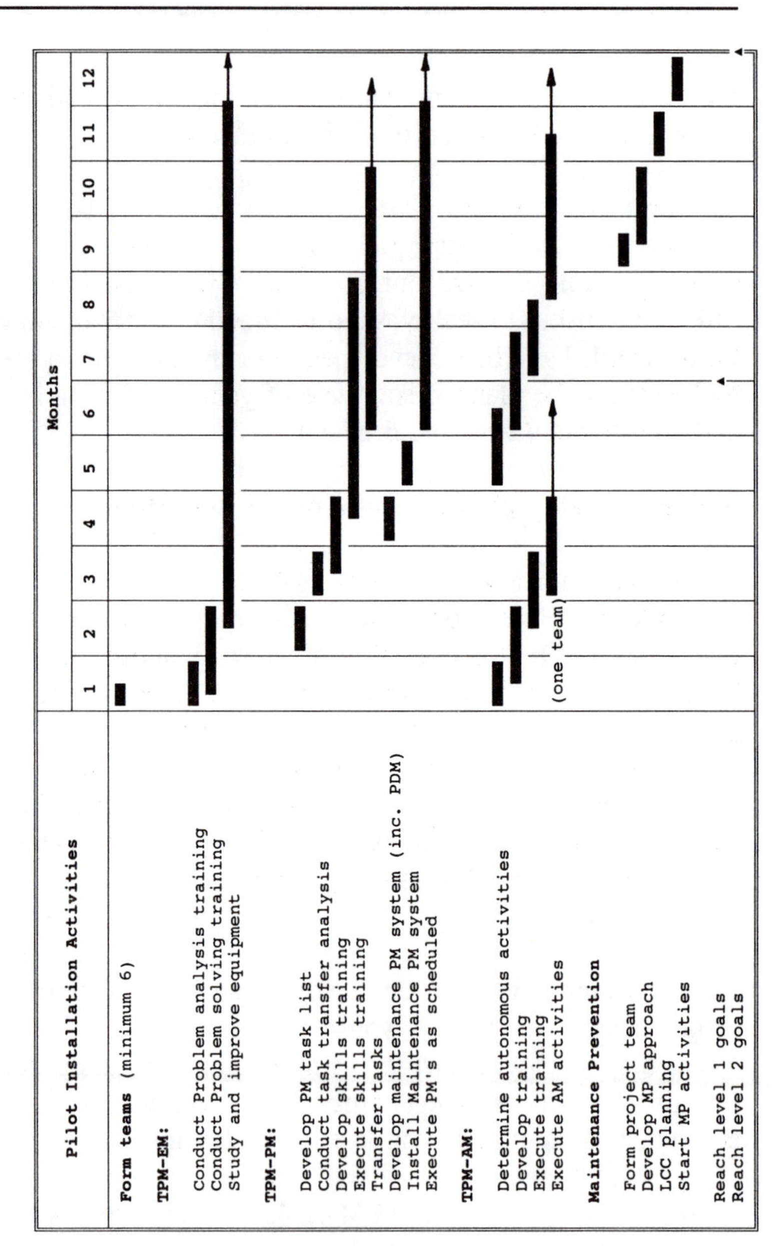

Figure 34

Routine cleaning and lubrication (including the development of procedures) comes next. This may involve training of the teams in the proper procedures. After that, develop inspection procedures for operators (with their participation) and begin the training process that will eventually lead the teams to autonomously inspect their equipment.

Within TPM-PM, plan to do a task transfer analysis first, based on existing (or to be developed) PM checklists. This analysis will determine which tasks can be transferred now and what type (and how much) skills training needs to be carried out. Time needs to be scheduled to develop this training and to execute operator PM training. The next phase of the schedule is the gradual transfer of PM tasks, as operators become capable and motivated to carry them out.

The development of the maintenance based PM system needs to be scheduled also. The results from the feasibility study will indicate how much work this will require. It may take a fair amount of time, if your PM system is not well developed. Follow the steps outlined in chapter VIII.

The development of effective Maintenance Management is normally not part of the pilot installation, as it concerns the whole maintenance department. Plan for it in the master plan and develop a separate schedule for this task if required.

However, it is advisable to think about MP (Maintenance Prevention) activities being part of the pilot. As the teams improve equipment and get involved in maintenance activities, the incentive rises to *prevent* maintenance in the first place and to focus on LCC (Life Cycle Costs).

As part of your pilot installation, you want to test all viable TPM options as early as possible to gain experience for your plant-wide installation. The pilot installation should stay about three months ahead of a phased total installation and will act as a "locomotive" for all other areas.

Step 9: Develop detailed installation plans

Although this is part of the installation planning, it is usually delayed until some data from the pilot installation is available. The approach is quite the same as for the pilot, unless the installation sequence changes. You should develop a separate plan for each area of installation and update the plans as required. Include details such as training schedules, meeting frequencies, goals and target dates. Additional details could be the sequence of machines or lines to be worked on, based on the needs determined from the feasibility study.

Determine the number of teams needed and develop, with the team leader, a work and training schedule for each team, based on the equipment needs and training needs of team members, as developed during the skills required/available analysis. Further training requirements will be determined during the PM task transfer analysis, done in the early part of TPM-PM installation. That means that you need to update the detailed installation plans from time to time.

The TPM office is responsible for maintaining all installation plans and coordinating all training. If there are limited resources for certain types of training, careful scheduling needs to be done. This office will also be responsible for administering the certification process and to keep each group's skills inventory, unless your training department wants to do it. The progress against plans will also be measured and reported by the TPM office for all areas.

Step 10: Present to management

Before the pilot installation is started, the available installation plans, especially the master plan and the pilot plan, should be presented to management or the Plant TPM steering committee. This will be the last meeting and final commitment before the actual installation starts. The occasion should be used to report on the TPM information, training and PR activities just completed and any feedback from this. The TPM vision, strategy, policies and goals should now be known to everybody in the plant and any problem or potential problem should be addressed. After this, your installation starts in earnest.

Phase II: Pilot Installation

The process for the pilot installation is quite the same as for the other areas of your plant-wide installation, except that it is the first and most important one.

Since you must succeed with your pilot installation, or at least have an excellent start, the selection of the area is very important. Choose an area or department that has the right "corporate climate;" in other words, people that are cooperative and eager to participate, wanting to demonstrate that they can improve their equipment and it's maintenance. The department should not be too large (between 50 and 100 people is ideal) and typically has had an excellent feasibility study, meaning that good data is available. Make sure that as many people as possible from the feasibility study team will be involved in the pilot installation, since they are already highly motivated (they have seen the problems and opportunities). They will be your "missionaries," motivating others to participate.

The purpose of the pilot installation is to "test" approaches before committing yourself to a fixed format for the total installation. Of course, you can't *complete* the whole pilot installation before starting your plant-wide installation, as you don't have that much time (two to three years!) to wait. But with an average three months head start, you can learn what works in your plant and what may need re-thinking. You can test a different sequence with different teams, such as most teams starting with TPM-EM and one or more teams starting with TPM-AM. In planning for the pilot, you can identify which machines need improvement and which equipment needs cleaning and other "autonomous maintenance" activities.

Form Teams

The first step is to create the teams (CATS). Hopefully you have created some teams during the feasibility study to execute the equipment condition analysis. Those teams will now be ready to continue. If not, extend all possible effort to organize a very high percentage (if not all) of your operators, maintenance crafts, engineers and supervisors into

TPM small groups.

Surprisingly, this is where a lot of companies have problems. There may be an existing team structure for other programs, or a "work group" concept that has vested interests. During the feasibility study, under "corporate culture", you should have picked up this situation and planned for it. It is sometimes possible to retain existing groups (such as manufacturing teams) or to split large work groups into smaller CATS. It may take the personal effort of the TPM Manager, or even the TPM Champion or another manager, to help organize the teams. It does not happen by itself. Without teams, you can't do TPM!

Each team should be led by a team leader. Again, the pilot installation offers a chance to test approaches. Use an engineer for one team, a maintenance supervisor and a production supervisor for others, a maintenance craft and a lead operator for others yet. There is no hard rule; you have to determine what works best in your situation. Obviously, they need to know how to lead a team and have sufficient technical knowledge to provide training and guidance for the team's work. Most teams like to give themselves a name, which you should encourage. Some even have their own logos.

Installing TPM-EM

In most non-Japanese plants, TPM-EM will be installed first. Once the teams are organized, you have to establish a meeting schedule. Since you want to proceed fairly fast with your pilot installation, weekly meetings of at least one hour duration should be your minimum. Most companies do better than that. Have your teams meet in a room in the plant that is equipped with a white board and a flip chart. One of the team members should be appointed secretary to take minutes.

If you have completed a good feasibility study, there will be plenty of data available to get started. Sometimes it is hard to decide *where* to start because there are so many pressing needs or opportunities. Review the OEE and loss analysis of the team's equipment first. Start with a significant loss, which is very often part of idling and minor stoppages or a frequent breakdown. Sometimes it is excessive set-up time.

If your team "owns" an assembly or production line of, let's say, six machines, the "Eight step method to establish equipment priorities for applying TPM to a production flow" can be used to determine where to start and why (see Figure 35).

Use the following approach:

Step 1: Establish the current actual output of each machine (i.e. pieces per hour) (CO)

Step 2: Determine your current OEE for each machine (COEE)

Step 3: Develop an attainable new OEE (NOEE) by studying the losses and estimating the impact of equipment and maintenance improvement. Surprisingly, this can be done with relative accuracy, if the losses are well-established and

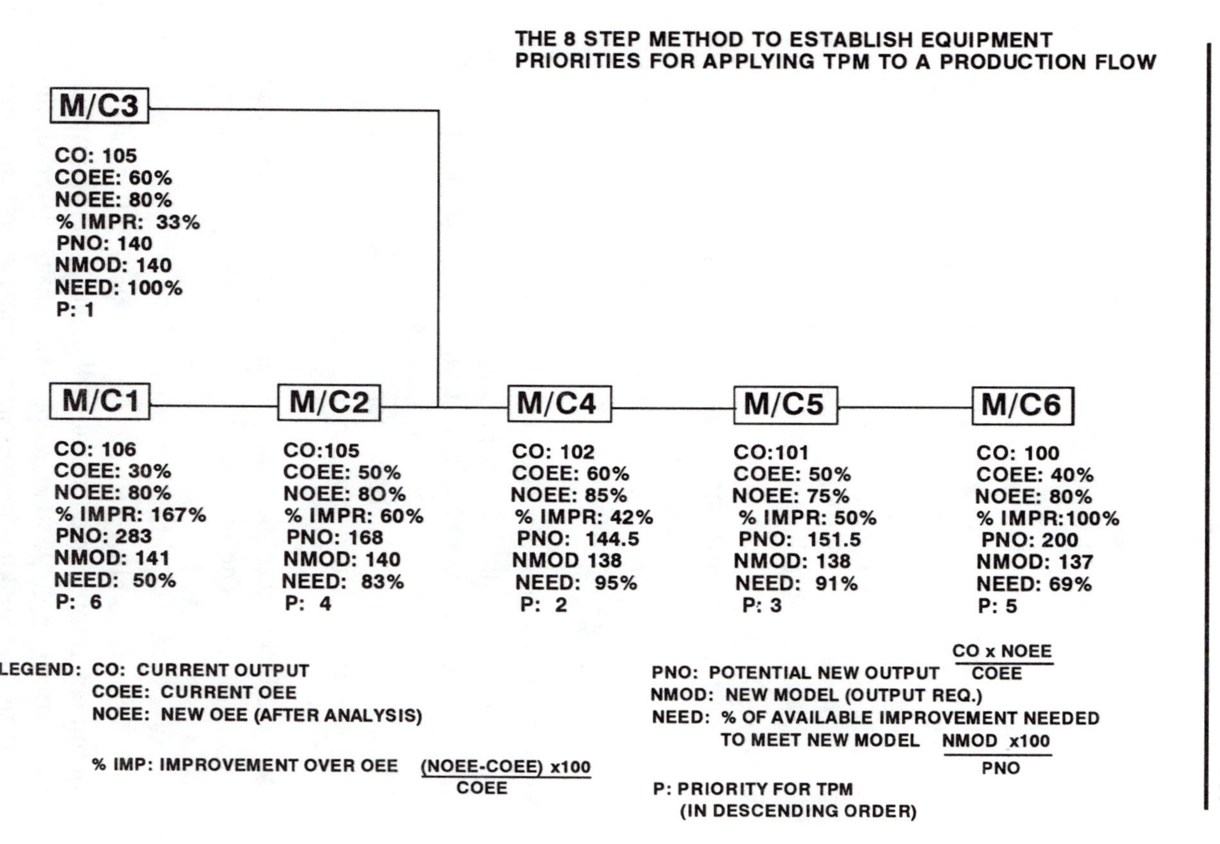
THE 8 STEP METHOD TO ESTABLISH EQUIPMENT
PRIORITIES FOR APPLYING TPM TO A PRODUCTION FLOW
M/C3
CO: 105
COEE: 60%
NOEE: 80%
% IMPR: 33%
PNO: 140
NMOD: 140
NEED: 100%
P: 1
M/C1
CO: 106
COEE: 30%
NOEE: 80%
% IMPR: 167%
PNO: 283
NMOD: 141
NEED: 50%
P: 6
M/C2
CO:105
COEE: 50%
NOEE: 80%
% IMPR: 60%
PNO: 168
NMOD: 140
NEED: 83%
P: 4
M/C4
CO: 102
COEE: 60%
NOEE: 85%
% IMPR: 42%
PNO: 144.5
NMOD 138
NEED: 95%
P: 2
M/C5
CO:101
COEE: 50%
NOEE: 75%
% IMPR: 50%
PNO: 151.5
NMOD: 138
NEED: 91%
P: 3
M/C6
CO: 100
COEE: 40%
NOEE: 80%
% IMPR:100%
PNO: 200
NMOD: 137
NEED: 69%
P: 5
LEGEND: CO: CURRENT OUTPUT
COEE: CURRENT OEE
NOEE: NEW OEE (AFTER ANALYSIS)
% IMP: IMPROVEMENT OVER OEE (NOEE-COEE) x100 / COEE
PNO: POTENTIAL NEW OUTPUT CO x NOEE / COEE
NMOD: NEW MODEL (OUTPUT REQ.)
NEED: % OF AVAILABLE IMPROVEMENT NEEDED
TO MEET NEW MODEL NMOD x100 / PNO
P: PRIORITY FOR TPM
(IN DESCENDING ORDER)

Figure 35

properly measured. This is a job for the CATS and will help to establish goals.

Step 4: Calculate the percent possible equipment improvement (% Impr.) by comparing NOEE to COEE.

Step 5: Convert your current output (CO) to the potential new output (PNO) for each machine

Step 6: Now you can establish a model (NMOD) for the new potential throughput of the improved line. Your constraint machine (M/C3 in the example) will dictate what the numbers for each machine should be for a balanced flow.

Step 7: This step will now determine what percentage of the total improvement potential for each machine will be needed (NEED) to bring this machine up to the number required for the new model. This number can vary widely, depending on the capacity of the machine and it's current OEE.

Step 8: Determine the priority for TPM (P) in descending order. The higher the percent need, the higher the priority.

This method allows you to focus your TPM activities on equipment improvements that will have a quick impact on your throughput. It also stops you from "overkilling" a machine, i.e. making improvements far beyond the point of need for the new model. Your resources, and the time for your teams, are limited, so apply them where it counts!

Eventually, you can go back to each machine to improve it to its "highest" potential, but if you are interested in an early break-even of your TPM program and a high ROI, the first go-around will improve your equipment to it's "highest *required*" potential.

Sometimes a pareto analysis is available from the feasibility study, but normally not. Train the team to develop these analyses, the first and second level pareto (see Figure 19). Assuming idling and minor stoppages was you biggest loss, and jams the largest percentage within that group, the team will now determine where the jams occur and quantify these, resulting in a third level pareto.

Now, the team will have to *define* the problem and describe it, including making sketches, etc. There are usually numerous reasons why jams occur. This activity will directly lead to *analysis* of the various reasons. Here, some discipline is required, as team members frequently jump to conclusions, since some solutions appear to be obvious. Remember, you need to *eliminate* the problems, not just make some temporary quick fixes.

By now, the team is usually heavily involved and you can sense excitement. But, again, you have arrived at another critical point. You must *execute* the *solutions* and *improvements* the teams have come up with, or you will kill their initiative quickly. Having maintenance participate in the CATS usually helps, but many projects include some costs and scheduled equipment downtime to install changes and improvements. You must follow through or your equipment improvement activities will be in jeopardy.

In chapter IX, numerous additional problem analysis and problem solving techniques are discussed. They include:

- OEE loss analysis (Fig.24)
- Equipment Condition Analysis (Fig. 20)
- Equipment history study (Fig. 17)
- Failure information sheet (Fig. 22)
- Condition-Action sheet (Fig. 21)
- Cause and effect diagram (Fig. 23)
- Root cause analysis
- Methods analysis

Your CATs will learn and apply these techniques as required in their work. Additional quality control and industrial engineering techniques may be useful. Involve these specialists and have them work with or train your TPM teams as required.

The team should remeasure the OEE of their equipment about every three to six months, more frequently if significant improvements have been made. It is important to measure and publicize progress made and results accomplished. They should calculate the benefits and include them in their reports. The TPM staff will accumulate results achieved by all teams and report to plant management or the steering committee. There are organizations, such as Ford Electronics Division (ELD), that require *monthly* OEE reports from all their plants.

Figure 36 shows an OEE Improvement Guide that can be used by the CATS as a checklist. It is a broad list of many possible actions addressing all major equipment losses. Sometimes teams get lost in details and it is necessary to

OEE Improvement Guide

OEE = Availability x Performance Efficiency x Rate of Quality

		✓
Availability	• **Reduce set-up time (major opportunity)** - Eliminate set-ups - Automate configuration changes - Reduce calibration time (automate) - Limit test runs (or during planned downtime) • **Eliminate breakdowns (major opportunity)** - Carry out equipment improvements (TPM-EM) - Improve PM (TPM-PM) - Introduce autonomous maintenance (TPM-AM)	✓
Performance Efficiency	• **Reduce idling/minor stoppages (major opportunity)** - Improve flow of materials - Change staffing (eliminate "no operator" loss) - Improve equipment (eliminate "jam" loss) (TPM-EM) - Introduce autonomous inspection (TPM-AM) - Introduce cleaning and lubrication (TPM-PM) • **Eliminate "Reduced Speed" loss (major opportunity)** - Replace worn parts (typially major PM) - Re-tighten all bolts - Balance all rotating components - Improve lubrication • **Introduce improved PM and Predictive Maintenance**	✓
Rate of Quality	• **Eliminate scrap and rework (major opportunity)** - Introduce SPC (Statistical Process Control) - Improve equipment adjustment (TPM-AM) - Introduce equipment monitoring (measure wear) - Establish tool replacement procedure (count hits/strokes,etc.) - Introduce autonomous inspection (TPM-AM) - Improve equipment (TPM-EM) - Improve cleaning and lubrication (TPM-AM/PM) • **Improve product quality (major opportunity)** - Maintain equipment accuracy (TPM-AM/PM) - Implement all actions above	✓

Figure 36

take a step back and review the total picture of opportunities.

This guide also serves another purpose. It points to necessary TPM-PM and TPM-AM activities which the team may not have thought about yet. Having made improvements on their equipment, the operators should now be much better motivated to participate in PM and AM activities to make sure their equipment stays in top condition.

Installing TPM-PM

Preventive maintenance is the single most important tool to keep your equipment in top running condition and to eliminate breakdowns. TPM-PM will bring your plant a giant step closer to this goal. This is the area where your operators can substantially contribute while at the same time reducing overall maintenance costs.

The challenge is to get your operators involved. If you have started with TPM-EM, they will most likely see the need for improved PM and will be better motivated to keep "their" machine in good condition. Now the question is, what can they do and how do you go about an orderly process of transferring appropriate tasks to them. You don't want to force your operators to do tasks they don't feel comfortable with, nor do you want to remove tasks from maintenance that they strongly feel they should be doing.

The best approach is to let the two parties figure that out themselves; in teamwork. Use the CATS to get the process started. Figure 37 shows a "PM task transfer analysis" that can be used as a vehicle. Take the existing (or future) PM

PM TASK TRANSFER ANALYSIS

Input from: TASK DESCRIPTION:	OPS. Operator wants to do task	MAINT. Maint. wants Operator to do	ASC* Operator can do task	REMARKS:
1.	√	√	√	Task can be transferred now
2.	√	√	no	Training needed first (identify)
3.	no	√		VETO by Operators (address later)
4.	√	no		VETO by Maintenance (address later)
5.				
6.				
7.				

*Area Steering Committee

	Ops.	Maint.	Superv.	Safety	Union
Skill	√	√	√	√	√
Safety	√	√	√	√	√

Figure 37

checklists for the team's equipment and for each task listed on the forms, ask three questions:

1. To the operators: Do you want to do this task?

2. To maintenance: Do you want the operators to do this task?

3. To both: Can the operator do this task?

There will be four possible outcomes:

a) A "yes" to all three questions means this task can be transferred now. However, before you do it, you must make sure that the operator has the skill to do it and *can do it safely*. Other parties, such as operations, maintenance, the area supervisor, the safety department, and sometimes the union (the area steering committee) need to sign off on that.

b) A "yes" to the first two, but a "no" to the third question means it is a task that can be transferred, but only *after training*.

c) The operator says "no" to the first question, for whatever reason. That's a veto and you let it go, at least for now.

d) The maintenance craft says "no" to the second question. It's also a veto and you let it go for now.

You may be surprised how much agreement, and therefore tasks to transfer there will be. If the operator is motivated, and maintenance recognizes they have better, more "high tech", things to do, it will happen. Now you must identify the training required for the b) items, develop the training and execute it. Very often, the maintenance personnel will do the training. Then the process under a) must be followed before the task is transferred.

Occasionally, there are tasks performed by operators that could be better done by maintenance. The task transfer process also works in reverse. Make sure the teams understand that this is not a one-way street and that "common sense" is the guiding principle.

After a few months, re-visit the items that have been vetoed. Usually, the operators fee comfortable with the tasks that they have been doing and are ready to take on more. Maintenance people recognize that the operators did well and agree to more tasks that they were not sure about before. Eventually, the operators will do all Type I PM tasks. Obviously, there are tasks that always will be done by maintenance, the Type II major PM functions.

You need to keep control over who is doing what, especially while the situation is fluid. At a certain point, for instance, on a Monday morning, the operators will take over certain tasks. Make sure that not only the checklists and procedures are available, but also the tools and materials, if any are required. Don't forget the reporting system that keeps control over completed tasks. One built-in control is still around, the maintenance personnel. If the operators fail to do the PMs right or completely, they will notice.

It is absolutely necessary to allow the operators *scheduled time* to complete their PM tasks. Therefore estimate the time requirement for the tasks and *shut the equipment down* for that period. This is very hard to do for an operations manager, especially if the equipment is in perfect running condition. But the reason why it is in perfect running condition, is that PM has been done regularly!

If properly planned, PM will only take a few minutes per day and production will gain a multiple of that amount in additional uptime. It's a good trade-off, but some production managers or supervisors really need to be sold on that!

The other part of TPM-PM are the Type II activities done by the maintenance department. With fewer routine PMs, the department is now in a much better position to plan and schedule these. Follow the ten step approach described in Chapter VIII.

Installing TPM-AM

If you have followed the sequence EM-PM, then TPM-AM (autonomous maintenance) will come quite naturally to the teams. There is a degree of overlap between PM and AM and sometimes the distinction is a bit blurred. However, there are specific tasks that are unique to autonomous maintenance, such as initial cleaning of equipment and actions based on initial cleaning. It is a more disciplined approach, relying on cleanliness, orderliness, organization and standardization, which may account for some of the resistance encountered in non-Japanese TPM installations that are following the Japanese model.

Once your teams are involved with the other TPEM components (EM and PM), the level of motivation should be high enough to go to TPM-AM. For the pilot installation, try to get one or more teams to start directly with AM, just to test the viability in your plant. Friendly competition with other teams may produce unexpected results.

Figure 38 shows a form that can be used to guide you through the initial cleaning of selected equipment or components of machines. Teams of operators and maintenance personnel start to thoroughly clean equipment. The activities need to be selected, time must be scheduled

TPM - AM		INITIAL CLEANING			EQUIPMENT________ NO.______					
EQUIPMENT COMPONENT	ACTIVITY	Name		Tools/ Materials	Schedule		Hrs. Used	Compl.	Notes	
		Op.	Tech		Start	End				
TPM TEAM ________			PREPARED BY: ________				DATE ________			

Figure 38

and tools and material need to be made available. Since the equipment needs to be taken out of service for prolonged periods of time, the planning steps are important.

Many surprises will show up during initial cleaning, such as lubrication points nobody knew where there, loose connections, bolts or wires, etc. It is quite a learning process for the operators. Since dozens, sometimes hundreds of problems or minor defects are discovered, there is an organized approach needed to first record and sort them, then to address them.

Here again, you can split the work into Type I activities that can be done by operators and Type II jobs that need to be done by the maintenance department. Some jobs can be done very well by a team of operators and maintenance, since the equipment is usually down when the corrections are made. Use a form as shown in Figure 39 to plan and schedule your actions based on initial cleaning.

Once the operators work on a sparkling clean machine, the incentive is the keep it that way. At this point, procedures should be developed for operator-executed cleaning and lubrication activities. Here is the similarity and overlap with TPM-PM. These tasks must be incorporated into existing or future PM task lists. As with PM, there may be training required before operators can execute all identified tasks.

The next logical step is for operators to inspect their equipment to look for deterioration and other problems. Typical areas for inspection are: oil levels; pressure gauges; function of moving parts, such as levers and switches;

TPM - AM

ACTION PLAN BASED ON INITIAL CLEANING

EQUIPMENT ____________ NO. ______

EQUIPMENT COMPONENT	PROBLEM FOUND	ACTION REQUIRED	Name		Schedule		Hrs. Used	Compl.	Notes
			Op.	Tech	Start	End			

TPM TEAM ____________ PREPARED BY: ____________ DATE ____________

Figure 39

hydraulic or pneumatic hose connections; bolting and other connections; wear of components; condition of safety items, such as guards and interlocks, etc.

Many of these activities will require training, before operators can do them autonomously. Determine your training tasks, develop the training and then execute the training. Very often, the CATS develop their own training and a lot of training material is transferrable to other teams. The TPM staff will coordinate and assist in the development of the material and assist in the scheduling and execution of the training, especially if outside resources are involved. Retired maintenance personnel is one of the possible outside resources.

Depending on the depth of the operator involvement in inspection activities, the training may take a long time. However, it is an excellent investment, as this type of operator activity results in vastly more reliable equipment that operates at a very high level of effectiveness.

When starting this activity, use a checklist as shown in Figure 40 to control the execution of the tasks. Obviously, the contents of such a checklist will grow as you make progress with TPM-AM. Note the designation of the skill level included in the checklist. As discussed in Chapter VIII, you should certify operators at various levels of skill and permit only properly certified operators to carry out the appropriate functions of equipment inspection.

Minor repairs offers another opportunity for operator participation through TPM-AM. As the example of replacing a rupture disk in Chapter IV illustrates, operators can take

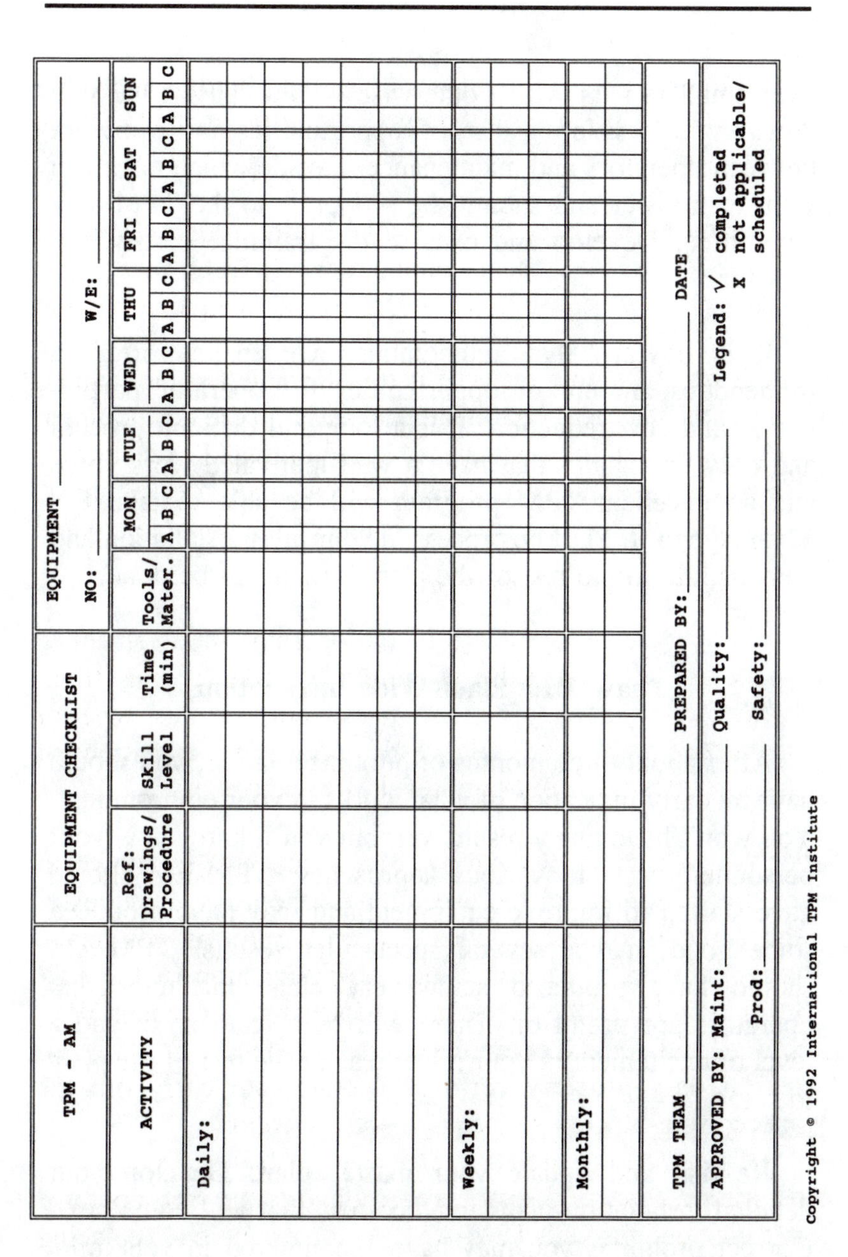

TPM - AM

EQUIPMENT CHECKLIST

EQUIPMENT ________ NO: ________ W/E: ________

ACTIVITY	Ref: Drawings/ Procedure	Skill Level	Time (min)	Tools/ Mater.	MON			TUE			WED			THU			FRI			SAT			SUN		
					A	B	C	A	B	C	A	B	C	A	B	C	A	B	C	A	B	C	A	B	C
Daily:																									
Weekly:																									
Monthly:																									

TPM TEAM ________ PREPARED BY: ________ DATE ________

APPROVED BY: Maint: ________ Quality: ________ Legend: √ completed

Prod: ________ Safety: ________ X not applicable/ scheduled

Figure 40

over small repairs which otherwise would cause excessive downtime. Develop a list of opportunities in teamwork between operators and maintenance. Decide which ones you want to transfer and submit the proposals to the TPM area committee. Develop and execute the training, certify and transfer.

It is obvious that autonomous maintenance offers a tremendous amount of opportunities for operator participation and cost reduction. It is a long and difficult process and only companies that have a well-motivated work force and an excellent TPM program will be able to take full advantage of it. Start he process in your pilot installation and demonstrate to the rest of the plant how it can be done.

Phase III: Plant-Wide Installation

After about three months of pilot installation, you should have an early indication of what works in your environment. You won't have many results yet, but you'll know how your personnel reacts to various approaches. TPM-EM should have started to improve equipment and may have produced some good, maybe several spectacular results. TPM-AM should have produced some very clean machines that operators are proud of. There will be a team spirit and a level of excitement in your pilot area for the rest of the plant to see.

Re-visit and update your master plan. Develop your detailed installation plans for the areas that will come next. Correct problems you may have encountered in your pilot installation and fine-tune your schedule. The approach and plans will not be much different from the pilot just discussed.

Form the TPM organization for the rest of the plant. Start with your area steering committees and develop as many CATS as you can. Typically there are not as many teams as you would like to have at the beginning, but it will grow. You are better off with fewer, but well-motivated and productive teams, than a large number of teams that don't do much. Word will get around and examples of improved and better performing equipment will be more and more.

Set equipment performance goals for all areas and challenge the area leadership to promote TPM. By now, you know which approaches your personnel will respond to. Publicize results and continue to expand into all areas as per plans. Be aware that TPM will take time. Most companies around the world that have an excellent TPM installation took at least three years.

Training

Operator skills training is critical to the success of your TPM installation. It is the most time consuming process of your installation. The TPM manager and staff must play a leading role in this area. This staff will develop the training manual and write the lesson plans, or assist the teams in doing so. Resources for the staff are the maintenance department and the training department. There are training companies that have skills training courses on video tape or interactive video disks in addition to lecture-based or self-study courses, using manuals.

However, most lesson plans will be developed in-house by the TPM staff and maintenance. They must include:

- Method
- Procedure
- Tools
- Materials
- Safety (OSHA)
- Sketches
- Pictures

There are two basic types of training: Classroom and on the job (OJT). The classroom will be used for more formalized training and when a white board, flip chart, computer or video needs to be used. However, hands-on training at the job site in frequent, but short lessons is used very effectively. Test what combination of approaches produces the best results for you. The need for certification at various levels of skill has been discussed earlier.

Maintenance Management System

Part of your plant-wide installation will likely include the improvement of your maintenance management system. Before TPM, your maintenance department may have been involved primarily with fire-fighting, which by nature does not involve much planning and scheduling. But now, with fewer breakdowns and less routine PM work, the type of work of the maintenance department will be major PM, predictive maintenance, equipment repairs, equipment improvement and overhauls, all of which must be *planned and scheduled*. Therefore, if not in existence, a formalized

Planning/Scheduling function should be established.

Also if not in existence, a computerized maintenance management system (CMMS) should be installed which must also support the operator-based PM and other maintenance activities. Bar code technology, as discussed in chapter VII, should be applied to all maintenance activities. In addition, a 52-week PM scheduling function is a must.

Reporting Progress

TPM must be a data-based system. Remember you are *managing* your equipment through TPEM, Total Productive Equipment Management. There are numerous measures, which typically are developed and reported by the TPM manager. These measures include:

- Progress of each area against plan
- Number of teams established and percent of employees involved
- Hours of training provided, including average by person
- Current skill level, by group and area
- Number of breakdowns and trend, by area
- Hours of downtime and trend, by area
- Current TEEP, OEE, and NEE, by equipment and area
- Current accumulated cost of TPM
- Current accumulated savings
- Current ROI
- Productivity improvements not included in savings
- Capacity expansion accomplished
- Other results that are included in the maintenance management system, such as MTBF, PM compliance,

maintenance productivity and utilization
- Other measures as may be requested by management

Recognition, Reinforcement and Celebration

A successful TPM installation depends on the commitment and positive attitude of everybody involved. Since TPM is a long cycle program, it is important to build and maintain this commitment and attitude, so it eventually becomes part of your corporate culture.

There are excellent, but often underutilized, tools to accomplish that. Recognition and positive reinforcement should be used by TPM and plant management in a conscious and organized fashion to create and maintain a high level of commitment and good attitudes.

Develop clearly defined *goals* or *milestones* for each team, such as:

- Amount of training (for instance 50 hours) or skill level to be reached
- Number of tasks completed (1000 level, 5000 level, etc.)
- Number of improvement projects completed, or amount of benefits accomplished
- A certain level of MTBF accomplished
- Attainment of a certain level of OEE

Then you must recognize and reward the accomplishment of the goals and milestones. There are many ways of doing that, and most are not expensive at all. They include: An article (with photos) in the plant or company newsletter;

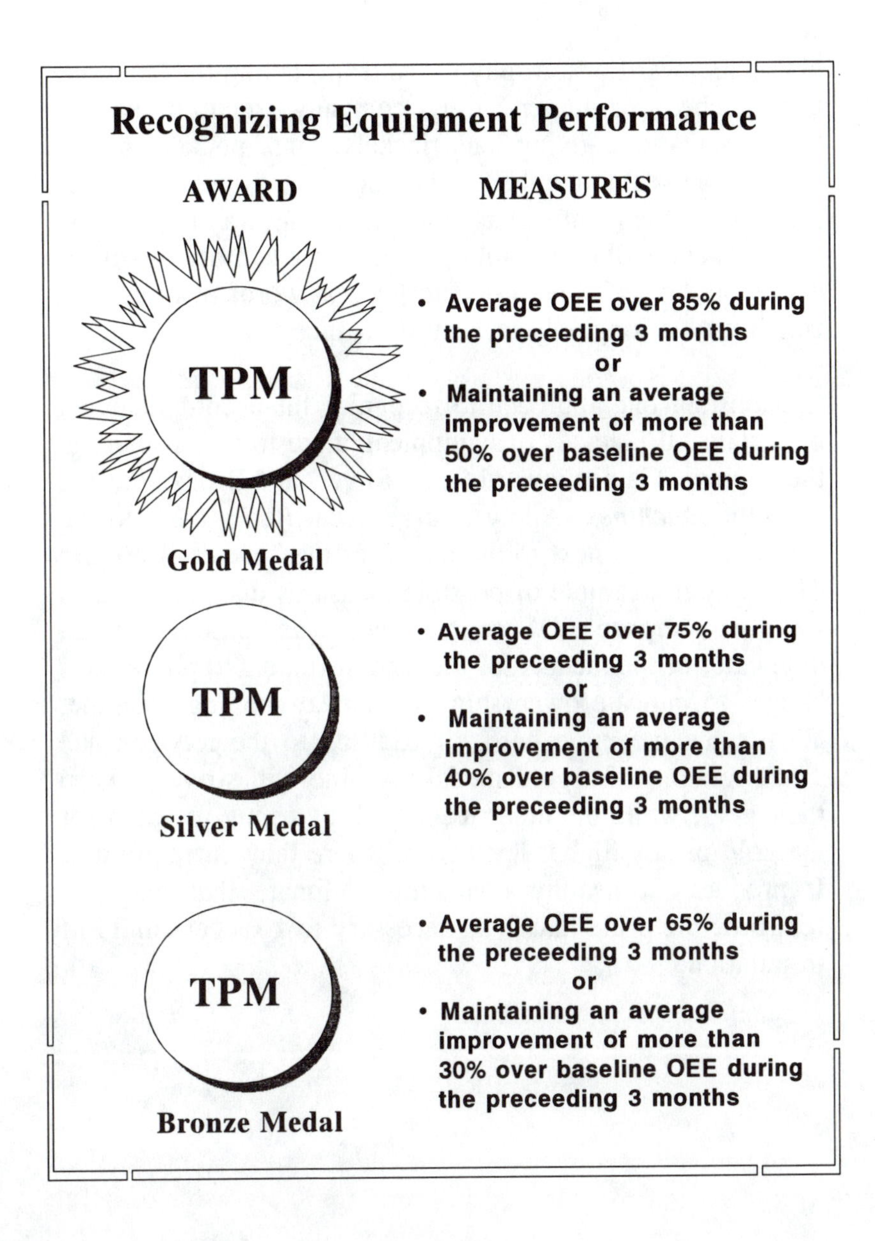
Recognizing Equipment Performance
AWARD
MEASURES
TPM
Gold Medal
• Average OEE over 85% during the preceeding 3 months
or
• Maintaining an average improvement of more than 50% over baseline OEE during the preceeding 3 months
TPM
Silver Medal
• Average OEE over 75% during the preceeding 3 months
or
• Maintaining an average improvement of more than 40% over baseline OEE during the preceeding 3 months
TPM
Bronze Medal
• Average OEE over 65% during the preceeding 3 months
or
• Maintaining an average improvement of more than 30% over baseline OEE during the preceeding 3 months

Figure 41

giving an award or a trophy to the team; taking the team out to a luncheon or dinner, with a company executive making the presentation; giving hats, jackets, pins, pocket knives, etc. to the team members; and more. A simple letter of praise from the plant manager will go a long way to motivate the team and will be proudly posted on the activity board for everybody to see. And, don't forget that words of praise are free but go a long way to keep up enthusiasm.

Consider one approach to recognize the accomplishment of certain OEE levels of equipment. Like to the winners at the Olympic Games, award Gold, Silver, and Bronze medals -- to the *machines*. Apply a large decal (the medal) to the equipment, right next to the names of the "owners." Figure 41 shows an example of possible measures that can be used for this purpose. Of course, you can produce small duplicates of the medals for the team members to pin to their badges to indicate ownership of a medal-winning machine. What's the incentive here, in addition to the recognition? The operators of a gold medal machine will strive to keep their gold, while the other teams will be trying to reach for the gold or any higher level from where they currently are. It produces a healthy competitive climate that helps to reinforce the pride that is so necessary to a successful TPM installation.

CHAPTER XII

"Fast-Track" TPM Installation

In many, especially American owned, companies around the world, there is considerable pressure to "go fast" with a TPM installation. There are several reasons for that. Often, a plant manager has a limited "tour of duty," in particular on overseas assignments, but also in large corporations, where rotation of management is commonplace. So results must be accomplished during his/her administration.

A much more important reason though is that the *need* for equipment improvement is so pressing, that any delay, or going slow, cannot be tolerated. There are plants that cannot produce the amount of goods that customers want and must turn away business. Often these plants have old equipment with low OEEs.

A third reason is that the typical Western manager just does not have the patience of his/her Oriental counterpart. No attempt shall be made here to discuss the pros or cons of that, but the fact remains that there is often the challenge to "fast-track" a TPM installation.

Can it be done? The answer is yes and no. You can accomplish results rather quickly by targeting specific equipment and focusing on it's early improvement through TPM-EM. Figure 42 shows an A-B-C analysis of TPM activities that can produce accelerated results. You have to determine what the A-activities are that will produce an early break-even or high ROI. You can go through the feasibility study, TPM training and installation planning quickly by assigning more resources to accomplish these functions.

International TPM Institute, Inc. offers a service called "TPM Blast-off," which is a one-week in-house accelerated TPM consulting and training program. The objective is to take a selected team of company employees (managers, engineers, maintenance and production employees) in a real industrial setting, from the initial TPM training to a custom-made TPM installation. The program includes the following phases:

- Intensive TPM and feasibility study training
- Inspection of equipment in a designated small pilot area
- Execution of the feasibility study in pilot area, including:

    -OEE observations and loss calculations on several machines
    -Equipment condition analysis, using teams
    -Skills available vs. skills needed analysis
    -Current and future maintenance analysis
    -Establishment of Type I and Type II PM tasks

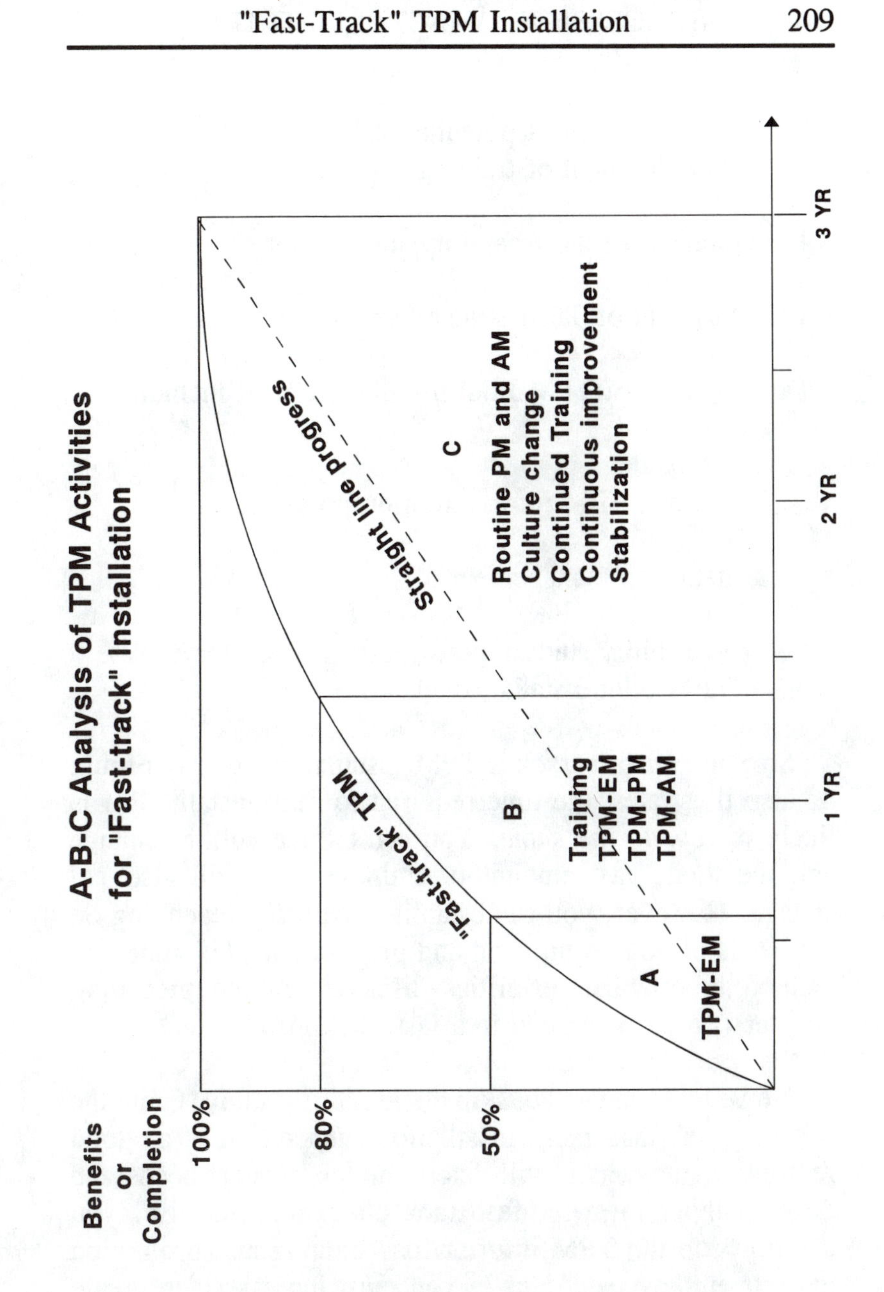
A-B-C Analysis of TPM Activities
for "Fast-track" Installation
Benefits
or
Completion
100%
80%
50%
Straight line progress
"Fast-track" TPM
A
TPM-EM
B
Training
TPM-EM
TPM-PM
TPM-AM
C
Routine PM and AM
Culture change
Continued Training
Continuous improvement
Stabilization
1 YR
2 YR
3 YR

Figure 42

-Development of potential task transfer tables
-Development of training needed

- Development of the feasibility study report

- Establishment of TPM strategy and goals

- Development of TPM Pilot installation plan, including:

  -Schedule
  -PR and TPM information proposals

- Presentation to management of:

  -Feasibility study report
  -TPM pilot installation plan

So you *can* fast-track a TPM installation, but as Figure 42 also illustrates, the time required to fully install TPM is likely to *remain the same*. You can't force culture change and the total, vast amount of training time will also not change. However, you can establish *priorities* depending on the needs of your equipment and production. The eight step method to establish priorities discussed in the preceding chapter is usually applied in a fast-track installation.

However, you can not skip the feasibility study. Quite the opposite, a fast-track installation *depends* on a good feasibility study, as it will determine the greatest needs and develop input for your fast-track planning. Likewise, you can not skip the TPM information, training and promotion to your employees. Actually you must increase it to create a higher level of readiness for a fast installation.

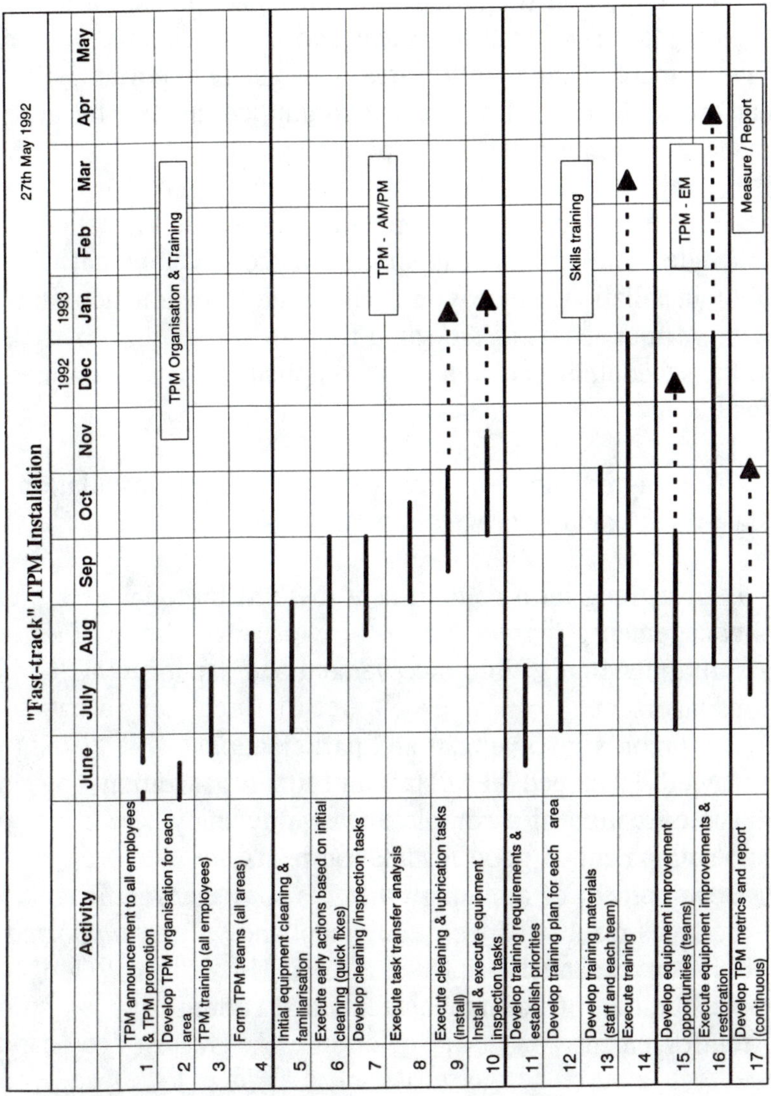
"Fast-track" TPM Installation
27th May 1992
1992
1993
Activity
June
July
Aug
Sep
Oct
Nov
Dec
Jan
Feb
Mar
Apr
May
1 TPM announcement to all employees & TPM promotion
2 Develop TPM organisation for each area
3 TPM training (all employees)
4 Form TPM teams (all areas)
5 Initial equipment cleaning & familiarisation
6 Execute early actions based on initial cleaning (quick fixes)
7 Develop cleaning /inspection tasks
8 Execute task transfer analysis
9 Execute cleaning & lubrication tasks (install)
10 Install & execute equipment inspection tasks
11 Develop training requirements & establish priorities
12 Develop training plan for each area
13 Develop training materials (staff and each team)
14 Execute training
15 Develop equipment improvement opportunities (teams)
16 Execute equipment improvements & restoration
17 Develop TPM metrics and report (continuous)
TPM Organisation & Training
TPM - AM/PM
Skills training
TPM - EM
Measure / Report

Figure 43

A strong, well supported TPM manager and staff are key ingredients in such an approach and the personal dynamic involvement of the plant or area manager is important. You can create the area TPM steering committees quickly because management can make the appointments.

Figure 43 shows an example of a "fast-track" installation schedule. The approach is similar to the development of a pilot installation, since you plan to start several activities with various teams at the same time. It is essential to stick to the schedule if you want to accomplish the results planned for.

The ingredients for any successful TPM installation, "fast-track" or not, are:

- A firm, long term commitment to TPM by top management
- Continued and visible involvement and support by management
- The union's involvement and participation
- A well developed TPM staff and line organization
- The execution of a complete feasibility study
- Development of good installation plans
- Development of a strategy that fits *your* environment, produces results for *your* equipment and will be *supported* by your personnel
- Recognition of accomplishments and continued reinforcement

If you follow these principles, you are virtually assured a successful and productive TPM installation.

About the Author

Edward H. Hartmann is the Founder and President/CEO of International TPM Institute, Inc. He was born in Switzerland and grew up in Zürich. After receiving his mechanical engineering degree, he emigrated to Canada and came to the United States in 1964. He studied industrial engineering at McGill University in Montreal and at the University of Southern California. Positions held in his early career were design engineer, industrial engineer and manufacturing engineer. He is a registered Professional Engineer (P.E.) in the State of California.

In 1969, he joined an international management consulting firm and eventually became a Senior Vice-President, member of the Board of Directors and President of its training division. Since 1986, he has studied the application and results of TPM during numerous plant visits to Japan and in 1987 formally introduced TPM to American industry at a TPM Executive Conference in Pittsburgh. Since that time, Mr. Hartmann has trained over 4,000 managers, engineers and other personnel in TPM.

International TPM Institute, Inc. offers training and consulting services in TPM and maintenance management worldwide. Its corporate offices are in the United States with regional offices in Santiago, Chile for South America, Zürich, Switzerland for Europe and Singapore for South East Asia.

Index

A-B-C analysis, 208
Asset utilization, 16, 17
AT&T, 1
Autonomous maintenance (AM) 34, 71, 159, 194, 198

Bar code technology, 106-107, 203
Baseline, 20, 64, 66, 148
Bottlenecks, 36, 130
Break-even point, 18, 33, 208
Breakdown maintenance, 11, 93, 143
Breakdowns 40, 42, 136, 145
Built-in diagnostics, 28

Cause and effect diagram (Fishbone diagram), 128-130
Changeover, 20, 29, 40
Chronic failure, 56
Cleaning, 21, 95, 142, 194, 196
Cleaning requirements, 23
Computerized maintenance management system (CMMS) 36, 113, 203
Constraint equipment, 12, 18
Control reports, 109
Corporate climate, 134, 181
Corporate culture, 76, 135, 146-147, 172, 204
Cost-benefit analysis, 20
Cost of operation, 26
Cost reduction benefits, 46
Cost reduction, 10-11, 43, 45
Creative action teams/ continuous improvement action teams (CATS), 116, 120, 127, 131, 164, 182-183, 201
Criticality, 100
Cycle time, 9-10, 33, 62-63
Cycle time reduction, 9-10

Debugging, 18, 31
Defect rate, 41
Department/area TPM committees, 164
Dai Nippon, 42, 45
Diminishing returns, 33
DiPasquale, Carl, 175
Downtime, 32-33, 40, 47, 148
DuPont, 1, 39

Eastman Kodak, 1, 39
Employee turnover, 142
Enfield/Treforest plants,

Enfield/Treforest plants (cont.), 173-174
Engineering specifications, 29
Environmental issues, 12
Equipment
 availability, 17, 55-56, 59, 62
 breakdown, 94, 97, 124
 cleaning, 22, 43
 condition, 25, 126
 condition analysis, 120, 126, 138
 cost, 26
 design, 26
 downtime, 32, 80, 187
 effectiveness, 15-16, 52-54, 59, 148
 failure(s), 21, 56, 100, 128
 history, 29, 105-107, 124
 improvement, 20, 75-77, 118, 128, 138, 186-187
 installation and debugging, 26, 31
  losses, 10, 20, 54-59, 136, 151
 management (EM), 17-18, 20, 23, 29, 69, 75-76, 115
 overhaul, 44, 94
 performance, 17-18, 35-36, 112, 136, 201
 productivity, 51-52
 speed, 32, 33
Equipment (cont.),
 utilization (EU), 17, 52, 59, 138

Failure information sheet (FISH), 124, 126, 154
Feasibility study, 18, 38, 63, 77-78, 133, 136, 208, 210
Feasibility study team, 149, 153
Five S's, 79, 159
Ford Electronics
 Division (ELD), 134, 173, 188
Ford Motor Company, 1, 39, 41, 133, 134, 147, 165
Ford's North Penn plant, 175

Gill, Andy, 173

Hidden losses, 57, 62
High tech, 43, 44, 192

IBM, 1
Idling, 10, 40, 42
Improved productivity, 42, 46, 67
Inspection, 21, 25, 159, 196, 198
Inspection procedures, 23, 179
Installation strategy, 158,

Installation stategy (cont.), 159, 175, 177
International TPM Institute, Inc., 16, 69, 208

Japan, 2, 3, 15, 71, 79, 133
Japanese, 2, 3, 4, 15, 16, 32, 39, 71, 79, 80, 134-135
Job satisfaction, 47
Just-in-time (JIT), 9, 82

Local area network (LAN), 113
Life cycle cost (LCC), 25-29
Line maintenance, 22, 80
Line organization, 37
Lubrication, 23, 41, 179

Maggard, Bill, 173
Maintenance,
- autonomous(AM), 34, 71, 76
- cost, 10-11, 42-43, 80-81
- management, 145-146, 179, 202-203
- prevention (MP), 179
- preventive, 34-35, 74, 94, 193

Major PM, 94-96
Management training, 169
Mean time between failures (MTBF), 20, 25, 108,
MTBF (cont.), 110
Methods analysis, 130
Motivation, 48, 90-91
Motorola, 1, 8, 39, 41, 147
"My machine" concept, 90, 96

Nakajima, Seiichi, 2, 3
Net operating time, 62, 63
Net equipment effectiveness (NEE), 54, 58-59, 64, 136
Nippondenso, 3
Non-Japanese culture, 15, 16, 34, 79

O'Connell, Michael, 134
OEE analysis, 120, 151, 184
OEE loss analysis, 118, 126, 184
On-the-job training (OJT), 84-85, 202
Operator certification, 88
Operator skills training, 201
Overall equipment effectiveness (OEE), 10, 16, 31, 52, 58, 184, 188

Pareto, first, second and third level, 118, 187
Pareto analysis, 118, 128, 187
Performance efficiency, 58, 62, 66

Pilot installation, 177-183
Planned downtime, 32-33, 59, 138
Planned maintenance, 32-33, 36, 59
Plant TPM steering committee, 162, 165, 181
Plant-wide installation, 180-182, 200-202
PM
 activities, 23
 checklist, 23, 98, 101
 compliance, 100, 107, 109
 costs, 110
 operator-based, 112, 203
 reports, 109-110
 requirements, 23
 task transfer analysis, 180, 190
Poling, Harold, 133, 165
Positive reinforcement, 131, 204
Predictive maintenance (PDM), 21, 44, 74, 75, 97
Preventive maintenance (PM), 2, 23, 25, 35, 74, 93-95, 190
Process defects, 54, 58
Procter & Gamble, 1
Production
 cost, 11
 equipment, 16
Production runs, 10, 43
Productivity, 18, 42
Progress reports, 110
Public relations, 158, 173

Q1, 147
Quality
 improvement, 18, 124
 rate of 58, 63-66

Rating scale, 122-124
Recognition, 204-206, 212
Reduced speed, 54, 57
Return on assets (ROA), 16, 52
Return on investment (ROI) 35, 68
Root cause analysis, 130
Routine maintenance, 79, 81-82
Routine PM, 95
Running time, 59

Safety, 45
Single minute exchange of die (SMED), 10
Six Sigma, 8, 147
Skill levels, 86, 88
Skills inventory, 88, 90, 181
Speed loss, 33, 40, 57, 62
Sporadic failure, 56
Statistical process control (SPC), 41, 82
Structured group interviews, 142
Szuluk, Charlie, 134

Taylor, Paul, 173
Team
 leader, 127, 150-151, 154, 180
 members, 118, 122, 131, 150-151, 153, 155
 size, 116, 149-150, 153
Teamwork, 47, 73, 122
Tennessee Eastman (TEC), 12, 45, 46
Texas Instruments, 1
Total effective equipment productivity (TEEP), 12, 52, 64
Total productive equipment management (TPEM), 16-18, 34-35, 69 -70
TPM
 activity board, 173
 -AM, 34-35, 71, 76-77, 83, 86, 90-91, 159, 177, 194, 198
 blast-off, 208
 champion, 37, 160, 162, 183
 coordinator, 37, 160
 development, 160-162, 177
 -EM, 34-35, 75, 115, 126, 183-184
 employee information, 170
 experience, 134
 goals, 31-33, 165, 168
 manager, 37-38, 160, 183, 201, 203, 212
 master plan, 175
TPM (cont.)
 -PM, 34-35, 74, 93-98, 179-180, 190-194
 small groups, 20, 37-38, 116, 122, 153, 164
 staff, 85, 162, 188, 198
Training, 71, 73, 82-85, 88, 141
 costs, 86
 levels of, 86
 requirements, 139, 141
 specific, 73
 time, 82, 85
Turnover rate, 142
Turnover, 48, 142
Type I activities, 23, 94-95, 97, 102, 196
Type II activities, 23, 94, 96-97, 102, 196
Types of PM, 94-95, 100, 102
Types of training, 202

Union(s), 5, 76-77, 112, 150
Unplanned (equipment) downtime, 32-33, 56, 59

Work orders, 102-103, 106-107, 113

Zero defects, 2, 31-32, 41

Also available from TPM Press:

"How to successfully install TPM in your Plant"
Eight-hour **video course** (1994 edition) for in-house training
by Edward H. Hartmann

Part I:
- Today's Manufacturing Issues and Challenges
- Equipment, the Focus of TPM (TPM Concept)
- The Power of TPM

Part II:
- How to measure your true Equipment Productivity
- Customizing your TPM Installation

Part III:
- Autonomous Maintenance (and how much do you need?)
- How to Design and Install an Effective PM Program
- Improving Equipment through Problem Solving Techniques

Part IV:
- The Feasibility Study
- The TPM Installation
- How to "Fast Track" a TPM Installation

Part I	$250.00	ISBN: 1-882258-02-9
Part II	$250.00	ISBN: 1-882258-03-7
Part III	$300.00	ISBN: 1-882258-04-5
Part IV	$350.00	ISBN: 1-882258-05 3
Complete set	$995.00	ISBN: 1-882258-06-1

Also available in Spanish at the same low prices
"Como instalar con exito el TPM en su empresa"